The Evolution of a Nation

THE PRINCETON ECONOMIC HISTORY OF THE WESTERN WORLD
Joel Mokyr, Editor

Growth in a Traditional Society: The French Countryside, 1450–1815, by Philip T. Hoffman

The Vanishing Irish: Households, Migration, and the Rural Economy in Ireland, 1850–1914, by Timothy W. Guinnane

Black '47 and Beyond: The Great Irish Famine in History, Economy, and Memory, by Cormac Ó Gráda

The Great Divergence: China, Europe, and the Making of the Modern World Economy, by Kenneth Pomeranz

The Big Problem of Small Change, by Thomas J. Sargent and François R. Velde

Farm to Factory: A Reinterpretation of the Soviet Industrial Revolution, by Robert C. Allen

Quarter Notes and Bank Notes: The Economics of Music Composition in the Eighteenth and Nineteenth Centuries, by F. M. Scherer

The Strictures of Inheritance: The Dutch Economy in the Nineteenth Century, by Jan Luiten van Zanden and Arthur van Riel

Understanding the Process of Economic Change, by Douglass C. North

Feeding the World: An Economic History of Agriculture, 1800–2000, by Giovanni Federico

Cultures Merging: A Historical and Economic Critique of Culture, by Eric L. Jones

The European Economy since 1945: Coordinated Capitalism and Beyond, by Barry Eichengreen

War, Wine, and Taxes: The Political Economy of Anglo-French Trade, 1689–1900, by John V. C. Nye

A Farewell to Alms: A Brief Economic History of the World, by Gregory Clark

Power and Plenty: Trade, War, and the World Economy in the Second Millennium, by Ronald Findlay and Kevin O'Rourke

Power over Peoples: Technology, Environments, and Western Imperialism, 1400 to the Present, by Daniel R. Headrick

Unsettled Account: The Evolution of Banking in the Industrialized World since 1800, by Richard S. Grossman

States of Credit: Size, Power, and the Development of European Polities, by David Stasavage

Creating Wine: The Emergence of a World Industry, 1840–1914, by James Simpson

The Evolution of a Nation: How Geography and Law Shaped the American States, by Daniel Berkowitz and Karen B. Clay

The Evolution of a Nation

HOW GEOGRAPHY AND LAW SHAPED THE AMERICAN STATES

Daniel Berkowitz and Karen B. Clay

PRINCETON UNIVERSITY PRESS

PRINCETON AND OXFORD

Published by Princeton University Press, 41 William Street, Princeton, New Jersey 08540
In the United Kingdom: Princeton University Press, 6 Oxford Street, Woodstock, Oxfordshire OX20 1TW

Library of Congress Cataloging-in-Publication Data

Berkowitz, Daniel.
The evolution of a nation : how geography and law shaped the American states / Daniel Berkowitz and Karen B. Clay.
p. cm.—(Princeton economic history of the western world)
Includes bibliographical references and index.
ISBN 978-0-691-13604-2 (hardcover : alk. paper) 1. U.S. states—Politics and government. 2. State governments—United States—History. 3. Law—United States—States—History. 4. U.S. states—Economic conditions. I. Clay, Karen B. II. Title.
JK2408.B469 2012
320.473—dc22

2011014585

British Library Cataloging-in-Publication Data is available

This book has been composed in Times

Printed on acid-free paper. ∞

press.princeton.edu

Printed in the United States of America

10 9 8 7 6 5 4 3 2 1

Dan: To Martha, Sam, Hannah, with love and thanks

Karen: To the Mowry men, Todd, Connor, Davis, and Grant

Contents

Acknowledgments

TWO EVENTS STARTED US DOWN THE PATH that has led to this book. The first was a long discussion with Ed Glaeser about his paper on legal origins in England and France during a lunch for the Carnegie Mellon University–University of Pittsburgh applied microeconomics workshop. The second was an equally long and stimulating discussion over breakfast with Stan Engerman where we discussed legal origins in the American states.

Our editor, Joel Mokyr, challenged us to write a book that the union of economists, historians, legal scholars, and political scientists can read. We hope we have risen to that level. Joel's careful reading and thoughtful prodding helped us clarify our arguments and presentation. Joel was aided by two anonymous reviewers who offered detailed and constructive comments on two complete drafts. While Joel and the reviewers worked much harder than anyone should expect, we take full responsibility for any of the shortcomings in this book.

We owe large personal and intellectual debts to scholars working at the intersection between economics, history, law, and politics, including Daron Acemoglu, Lance Davis, Stan Engerman, Avner Greif, Doug North, Katharina Pistor, Gerard Roland, Andrei Shleifer, Ken Sokoloff, and John Wallis. This project has benefited from the many comments we received from James Anderson, Martha Banwell, Ed Berkowitz, Hannah Berkowitz, Sam Berkowitz, Hans Bernd-Schafer, Dan Bogart, Eric Brousseau, Mehmet Caner, Ken Chan, Chris Connelly, Patrick Conway, Vivian Curran, Dave DeJong, Melissa Dell, Ying Fang, Lawrence Friedman, Robert Gordon, Amalia Kessler, Dan Klerman, George Krause, Todd Mowry, Baozhi Qu, Sam Rittenberg, Mel Stephens, Robert Strauss, Werner Troesken, and Randy Walsh.

We have also greatly benefited from the comments of seminar participants at the National Bureau for Economic Research workshop on the American economy, 2003; the Heinz School at Carnegie Mellon, 2003; the Indiana University Economic History Seminar, 2004; Carnegie Mellon's Political Economy reading group, 2004; the Allied Social Science Association Meetings in 2005 and 2008; the International Society for New Institutional Economics, 2006 and 2009; the American Law and Economics Association Meetings, 2007; the Christian A. Johnson Economic Enrichment Fund Lecture at Middlebury College, 2007; North Carolina State University, 2007; University of North Carolina at Chapel Hill, 2007; the Workshop on Law and Institutions at the School of

Oriental and Asian Studies of the University of London, 2007; the Law and Institutions workshop sponsored by University of Paris in 2007; the University of Pittsburgh Law School, 2008; Xiamen University, 2008; the European Society for New Institutional Economics, 2008; Tsinghua University, 2008; Lingnan University, 2008; the Chinese University of Hong Kong and City University of Hong Kong Joint Symposium on Institutions; Finance and Economic Development, 2009; the City University of Hong Kong, 2009; the Chinese University of Hong Kong, 2009; University of California, Irvine, 2009; Stanford Law and Economics Lunch, 2009; Economic History Association, 2009; and the Conference on Empirical Legal Studies, 2009. Karen Clay thanks Stanford Law School for its hospitality and support during the 2008–2009 academic year. Daniel Berkowitz thanks the City University of Hong Kong for its hospitality and support during the fall of 2009.

This book would not have been possible without a number of people who provided practical assistance. John Curry of the Census Bureau spent a great deal of time explaining state government budget data, and Eileen Kopchik spent many hours helping us download, merge, and understand the many data sets we have used. Lauree Graham spent many hours assembling data sets and typing in text and data. Amanda Gregg and Jeff Lingwall also provided capable research assistance, and Gary Franko did yeoman's work on the maps. At various points, Oliver Davies, Rose Goff, Gerrie Halloran, Gretchen Hunter, Joseph Plummer, and Elizabeth Thomas provided capable editorial assistance. Our production editor, Lauren Lepow, was indispensable in bringing this project to completion.

The Evolution of a Nation

CHAPTER 1

Introduction

COUNTRIES AROUND THE WORLD exhibit striking differences in per capita income. For example, in 2008, income in the United States, Singapore, and Switzerland was roughly forty times higher than income in Nepal and Uganda. There are also differences within countries. In the United States in 2000 income in the state of Connecticut was almost twice as high as income in the state of Mississippi. In Russia, income in the city of Moscow was six and a half times higher than income in the neighboring Ivanovo oblast.[1] What drives the disparities?

The disparities appear to be driven in part by political and legal institutions.[2] Political institutions such as legislatures influence key aspects of the economy, including the rights individuals hold vis-à-vis land, labor, capital, materials, and intellectual property. Legal institutions—in particular courts—play an integral role in defining and enforcing rights. This discussion pushes the question back one level. What drives differences in political and legal institutions across countries?

A recent literature suggests that differences in institutions and income are driven by a combination of geographic and historical factors. Gallup, Sachs, and Melllinger (1998), Mellinger, Gallup, and Sachs (2000), Sachs and Malaney (2002), Sachs (2003), and Nunn and Puga (2009) argue that geography influences income through its effects on public health, productivity, trade costs, population growth, and investment. Diamond (1997) makes the case that distance from historically critical trade routes and centers of knowledge influences income through its impact on the diffusion of technology and knowledge.

Several studies argue that geography and the disease environment at time of settlement have influenced the character of institutions in former European colonies. Engerman and Sokoloff (1997 and 2005) argue that climate and soil

[1] For countries, the data are gross national income adjusted for purchasing power parity from the World Bank. For the American states, the data are from the U.S. Census Bureau. For the Russian regions, the data are from various sources in the Russian statistical agencies: see Berkowitz and DeJong (2011).

[2] This work owes a debt to earlier studies of institutions, notably, North (1966), Davis and North (1971), North (1981), North (1990), Ostrom (1990) and Greif (2006).

shaped the subsequent character of political institutions. In colonies that were warm and rainy and had soil suitable for sugar and other staples, "bad" political institutions representing the narrow interests of wealthy elites emerged. In colonies that were colder and dryer and had different soil conditions, "good" political institutions representing broader interests were established. Acemoglu, Johnson, and Robinson (2001) provide evidence that the disease environment at time of settlement shaped the quality of institutions that protect property rights. In colonies where early settlers had a good chance of surviving, "good" institutions that protected property rights and limited the power of the government to expropriate emerged. By contrast, in colonies where early settlers were likely to contract life-threatening diseases, "bad" institutions that allowed settlers to easily extract resources emerged.

Historical factors such as legal and governmental institutions also appear to have been influential.[3] France and many other European countries inherited or appropriated a civil-law legal system early in their histories. Although civil law is conventionally referred to as a legal system, it represents a particular approach to governance that goes well beyond the courts.[4] Through colonization, these countries spread civil-law legal systems to many other parts of the world, including North America, South America, Asia, and Africa. England, for a whole host of historical reasons, developed a quite different legal and governmental system that came to be known as common law. Through colonization, it too spread common law to many other parts of the world.

Documenting how and why geography and other historical factors have had a persistent influence on political and legal institutions is challenging. The challenge arises because many countries lack the detailed qualitative and quantitative evidence necessary to document persistence and to test the relevance of alternative mechanisms. Lacking data on political institutions, Sokoloff and Engerman (2000) investigate a variety of indirect measures such as the timing and intensity of the extension of the voting franchise, the funding of public schools, and the allocation of land grants to immigrants in the Americas. Acemoglu and Robinson (2006) use a model to explain persistence of political institutions even in the face of large changes in the franchise. Glaeser and Shleifer (2002) and Klerman and Mahoney (2007) use historical evidence on England and France to show how legal origins shaped the evolution of legal

[3] Coatsworth (1993), Easterly (2006), Engerman, Mariscal, and Sokoloff (1998), Levine (2005), and Young (1994) describe political institutions that were created by European settlers and endured after colonization.

[4] See LaPorta et al.'s (2008) survey article.

procedures and judicial independence.[5] Our attempts in two earlier papers to understand how and why colonial legal institutions have had persistent effects on American state constitutions and state courts (Berkowitz and Clay 2005, 2006) were a major motivation for this book.

This book uses detailed historical evidence to analyze how and why geographical and colonial initial conditions have affected the evolution of legislatures and courts in the American states.[6] The American states have relatively diverse geographic and colonial initial conditions, well-documented historical experiences, and rich data on politics and courts going back to the 1860s. At the same time, a focus on the American experience avoids the problem of analyzing countries that often differ along many different dimensions and have had wildly different historical experiences. The primary goal of this book, then, is to understand political and legal institutions. In the conclusion, however, initial conditions are used to shed light on the contribution of political and legal institutions to long-term growth.

Figure 1.1 outlines the structure of the argument in the book. It is useful to begin by considering the two types of institutions of interest—state legislatures and state courts—near the top of the figure. Political competition in state legislatures is of interest because it is thought to lead to better economic and social outcomes. In the international context, Gryzmala-Buesse (2007), Jackson et al. (2005), Rodrik (1999), and Remington (2010) have found strong positive associations between the extent of political competition and outcomes such as government efficiency and corruption, the entry and subsequent growth of new firms, the provision of public goods, tax compliance, and manufacturing wages. The relationship between political competition and economic and social outcomes in the United States has been the focus of considerable discussion, but causal inference has been difficult. Besley, Persson, and Sturm (2010) use the 1965 Voting Rights Act as a source of exogenous variation—the federal government forced many southern states to allow registration of practically all individuals of voting age. They show that political competition was associated with growth through its influence on probusiness policies such as lower state taxes, higher state infrastructure spending, and the increased likelihood of a state having a right-to-work law.

[5] See also Banerjee and Iyer (2005), Iyer (2010), and Dell (2009).

[6] The analysis focuses on the forty-eight continental states. Alaska and Hawaii are not geographically contiguous, entered the union much later, and have had very different histories than the other states.

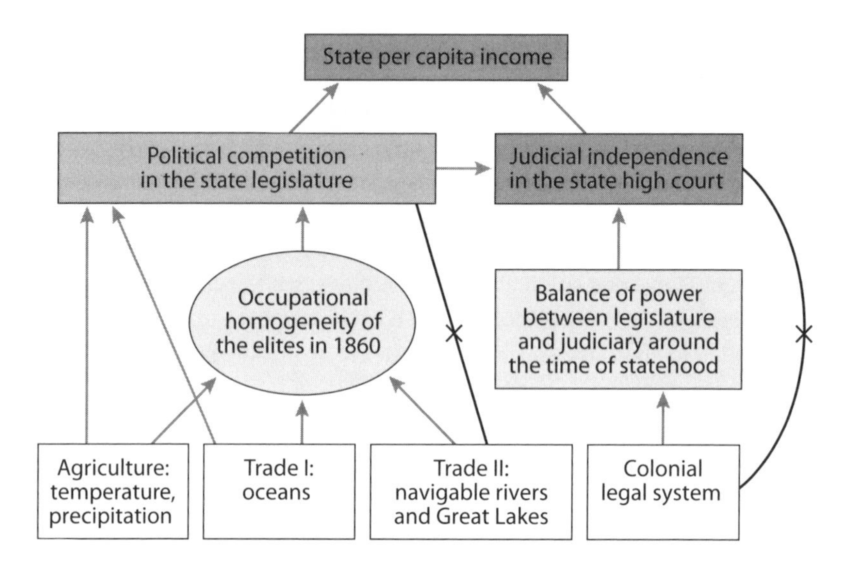

Figure 1.1. Outline of the Argument.

Judicial independence in the state high court is also related to important political and economic outcomes. Using a large sample of countries, La Porta et al. (2004) have shown that judicial independence is associated with stronger security of property rights, lighter government regulation, less state ownership, and more political freedom. A determinant of judicial independence in the American states is whether judges are elected.[7] Former U.S. Supreme Court justice Sandra Day O'Connor has warned of the threat to judicial independence created by the "flood of money into courtrooms by way of increasingly expensive and volatile judicial elections."[8] Moreover, the available evidence suggests that courts in states where sitting judges face partisan elections issue higher tort awards, rule more frequently against out-of-state businesses, have a higher likelihood of siding with state agencies in challenges to regulations, have a lower likelihood of enforcing constitutional restrictions on deficit financing, and also have more punitive sentencing outcomes.[9]

[7] The United States because it is the only country that allows high (state) level judges to be elected.

[8] Carey (2009).

[9] See Besley and Payne (2003), Tabarrok and Helland (1999), Hanssen (1999), Bohn and Inman (1996), and Huber and Gordon (2004).

Levels of interparty competition in state legislatures and the levels of independence of judges in the state's highest courts have been highly persistent over the period 1866–2000. Interparty competition is measured by examining the division of seats between the Democrats and the Whigs and later between the Democrats and the Republicans. This division is measured by the Ranney index of political competition. The index runs from 0, when one party holds all of the seats, to 100, when the parties each hold the same number of seats. The level of independence of judges is measured on a nine-point scale that captures what a state high court judge needs to do to remain on the bench. Having to run for reelection in a partisan race is considered the lowest level of independence, because judges may feel pressure to make politically popular decisions, even if they consider the decision to be legally incorrect. Having a lifetime appointment is considered the highest level of independence, because judges can make whatever decisions they believe are correct with virtually no political ramifications.

Figures 1.2 and 1.3 demonstrate the extent of this persistence of interparty competition and judicial independence. Persistence is measured by examining the correlation in political competition or judicial independence over time. If relative levels are persistent—states with high levels of political competition or judicial independence had high levels in other time periods—then the correlation between time periods should be high. Conversely, if they are not particularly persistent, then the correlation will be low. Figure 1.2 shows that the political competition in state legislatures in 1900–1918 was quite highly correlated with other subperiods during 1880–2000. The period 1866–1878 differs, primarily because many southern states had relatively high levels of competition under Reconstruction. Figure 1.3 presents an analogous figure for judicial independence in the state's high court. The high correlation of judicial independence in 1900–1918 with all of the other subperiods is striking.

The fact that levels of competition in state legislatures and independence of judges on state high courts are so persistent is surprising given the many changes that have occurred over the historical period 1866–2000. Population growth, immigration, urbanization, internal migration, the development of manufacturing, wars, the Great Depression, and the New Deal are only a partial list of the changes. Yet apparently these changes had limited effects on political competition in state legislatures and the independence of judges on state high courts.

One outcome of particular salience to many policymakers is per capita income. State legislatures and state courts are believed to shape per capita income. Per capita income, like relative levels of political competition in the state

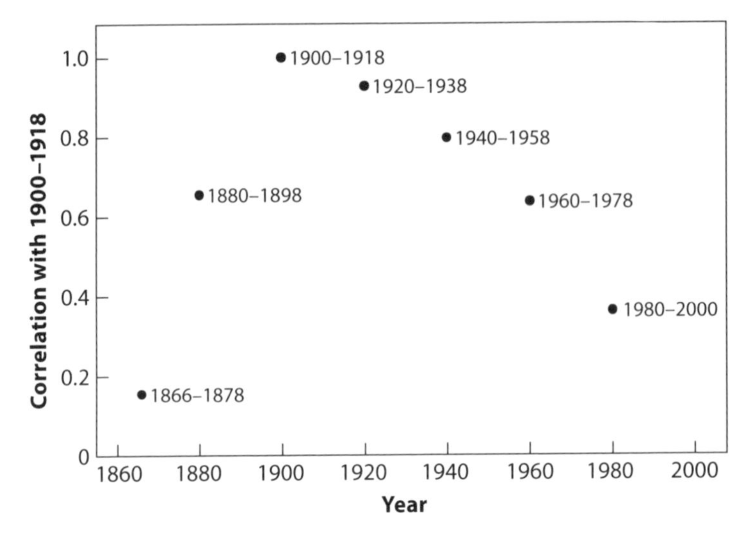

Figure 1.2. Persistence of Political Competition in State Legislatures, 1866–2000.
The Ranney index is used as a measure of political competition. Its construction is described in chapter 3. The Ranney index runs from 0 (no political competition) to 100 (highest possible political competition). Because Nebraska had a unicameral legislature for most of 1866–2000, it is not possible to measure its Ranney index. Thus, Nebraska is dropped from the sample. Louisiana is dropped because it kept a civil-law system after entering the union. Eleven additional states are dropped for lack of data. This leaves 35 states in the sample. The results are similar if we include these 11 states and conduct the analysis for 1910–2000.

legislature and the independence of judges on the state high court, is highly persistent. Figure 1.4 plots the correlation of per capita income in 1900 with six other years from 1880 to 2000. Although our primary focus is on state political and legal institutions, the last chapter briefly examines their influence on state per capita income.

The high degree of persistence suggests that conditions early in a state's history may have played a formative role in shaping political and legal institutions. The left side of figure 1.1 outlines the initial conditions that we argue shaped political competition in state legislatures and the mechanism through which the initial conditions acted on the legislature. The initial conditions represent state endowments that help determine a state's suitability for agriculture and for trade. States with moderate or warm temperatures and higher levels of precipitation were generally better suited for agriculture than states with

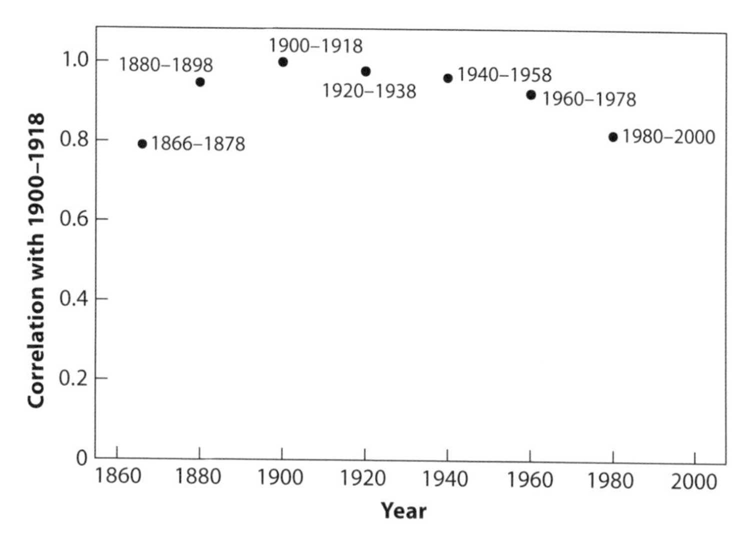

Figure 1.3. Persistence of State High Courts' Judicial Independence, 1866–2000.
The judicial independence index runs from 1 (partisan elections and least independent) to 9 (life time tenure and most independent). This index was constructed by Epstein, Knight, and Shvetsova (2002) and is discussed in detail in chapter 5. Louisiana is dropped because it kept a civil-law system after entering the union. For consistency with the previous figure, Nebraska is dropped because it had a unicameral legislature. Eleven additional states are dropped for lack of data. This leaves 35 states in the sample. The results are similar if we include these 11 states and conduct the analysis for 1910–2000.

cooler temperatures or low levels of precipitation. Similarly, states that were relatively close to the ocean and to internal water sources such as navigable rivers and the Great Lakes were better suited to trade than states that had more limited access to water transportation.

The intuition that initial conditions related to agriculture and trade may have shaped political competition is not especially novel. What is novel is that this book establishes a mechanism through which agriculture and trade acted on political competition in the state legislature. To understand the mechanism, one has to understand how seats in state legislatures were allocated. For most of the nineteenth and twentieth centuries, seats in state legislatures were allocated on the basis of geographic units such as counties and not population. Counties typically had a comparative advantage in either agriculture or trade. Thus the wealth of local elites was typically grounded in one of these two areas. Local

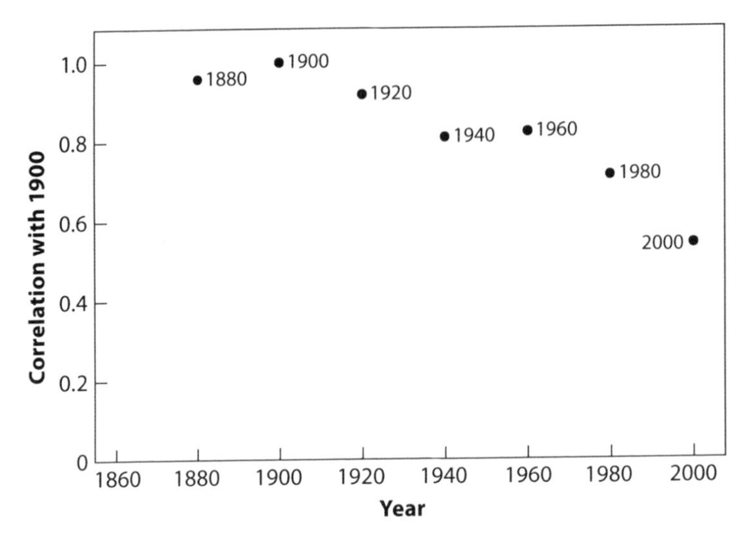

Figure 1.4. Persistence of State Per Capita Income, 1880–2000.

In 1880 there are data for 45 states of the 48 continental states. In all other years there are data for all 48 continental states.

elites tended to control who served in the state legislature. The two types of elites tended to have different interests and thus support different political parties. Economic activities in counties tended to change slowly over time. So an agricultural county tended to have agricultural elites who tended to send individuals with the same party affiliation to the state legislature. Similar trajectories occurred in counties with merchants or other types of elites. States with more occupationally diverse elites had higher levels of political competition in the state legislature than states where the elites were more homogeneous.

The wealth of the elite has been offered as a mechanism through which initial conditions might shape political competition.[10] States with wealthier elites would have more limited political competition, because the elites would more fully control politics. States with less wealthy elites would have greater political competition. These elites would choose not to devote resources to controlling politics or, if they did devote resources, would be less effective at controlling competition.

[10]This link is explicit in Engerman and Sokoloff (1997 and 2005), and implicit in Acemoglu, Johnson, and Robinson (2001) and much of the literature on the South.

Using data from the 1860 Census of Population and data on state political competition for 1866–2000, the relationships between occupational homogeneity and political competition and wealth and political competition are investigated. Occupational homogeneity of the elite in 1860 was strongly negatively related to political competition from 1866 through the end of the 1970s. In contrast, wealth of the elite had a variable relationship with political competition. In most periods the effect was small and positive, but in two periods it was large and negative.

We argue that the occupational homogeneity caused political competition. As the arrows in figure 1.1 suggest, temperature, precipitation, and distance to an ocean could in principle influence political competition in the state legislature through other channels. The distance to internal water transportation—navigable rivers and the Great Lakes—on the other hand, became much less important with the rise of the railroad. This change was sufficiently dramatic that internal water transportation was arguably only acting on political competition through the occupational homogeneity of the elite. Thus, using internal water transportation as a source of exogenous variation, occupational homogeneity of the elites is shown to have had a causal effect on political competition in state legislatures during 1866–1978 and, in some specifications, during 1866–2000.

The influence of the occupational homogeneity of the elites in 1860 persisted for interrelated reasons. Economic activities in counties tended to change slowly over time. As the underlying mix of economic activities changed, the mix of elites changed, but it took some time for new elites to grow powerful enough to elect individuals with other political affiliations. Persistence was greatly aided by the fact that geographic units were rarely reapportioned to reflect shifts in population. When combined, economic and political factors created strong persistence in the political composition of the state legislature.

One question is how slavery and the Civil War fit into the preceding discussion. Controlling for slavery does not substantially alter our results. Occupational homogeneity remains strongly negatively related to levels of political competition in the state legislature from 1866–1978, although the magnitude of the negative effect is smaller than it is without controls for slavery. Clearly slavery and the Civil War had an important influence on American political history. Few, if any, scholars would argue that it did not. Our point is that occupational homogeneity had an important influence on political competition in state legislatures above and beyond slavery. Some northern states, including Vermont, Indiana, Iowa, Kentucky, and New Hampshire, had high levels of

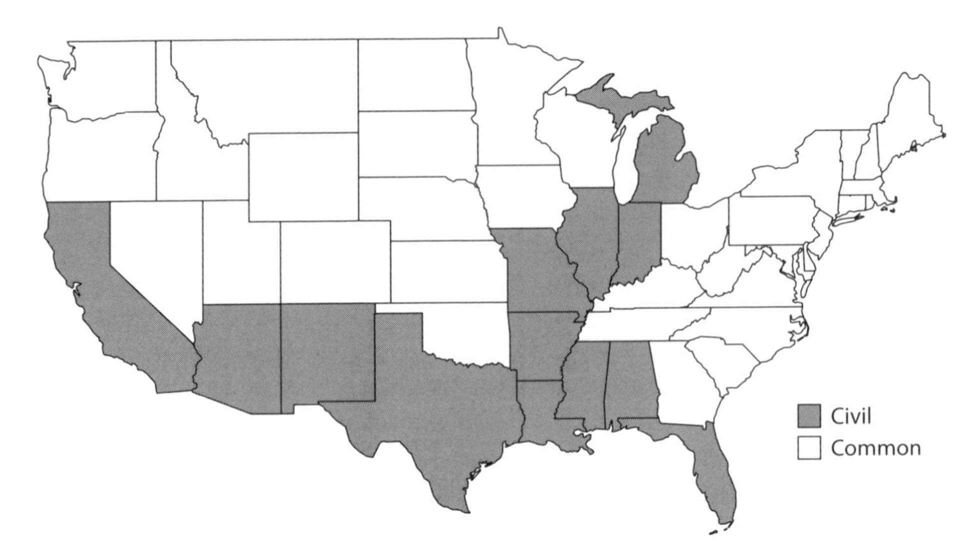

Figure 1.5. Civil-Law and Common-Law States.

The terms "civil law" and "common law" refer to states' colonial legal systems. All of the civil-law states except Louisiana ultimately adopted common law. The details of the classification of states are presented in chapter 2.

occupational homogeneity and low political competition over time. And some southern states, including Tennessee, Virginia, North Carolina, and Louisiana, had lower levels of occupational homogeneity and higher levels of political competition over time.

Turning to the right of side of figure 1.1, we argue that the colonial legal system was an initial condition that shaped courts. Many American states were first settled by European countries other than England. Thirteen states had operational civil-law legal systems at some point after 1750. Figure 1.5 shows the location of these states. All but one of these states, Louisiana, would adopt common law around the time of statehood. By the late eighteenth century, civil-law and common-law legal systems differed in many respects. One important respect in which they differed was the balance of power between the legislature and the courts. In common-law systems, legislatures and courts were relatively more equal. In civil-law systems, the legislature was relatively dominant and the judiciary subordinate. If civil law had a persistent affect, despite the adoption of common law everywhere except Louisiana, judges should be less independent in civil-law states.

For the twelve states that adopted common law after having had civil law, the influence of civil law comes through the balance of power and not through other channels. In particular, the influence is not coming through the survival of civil-law laws or procedures. Lawrence Friedman (1986) argues persuasively that, with a few exceptions related to property and family law, the common law completely "obliterated" civil law. Unfortunately, it is not possible to directly observe the balance of power. The historical record shows that individuals with civil-law backgrounds were active in early legislatures and constitutional conventions. Individuals with common-law backgrounds may have preferred a more dominant legislature or subordinate judiciary for other reasons, including the negative effects of the transition in legal systems on the courts or perceived benefits to having a balance of power that differed from the balance of power in most common-law states. Whatever the circumstances, the balance of power between state legislatures and courts appears to have been established during this early period. The reasons for persistence are less clear than they are for politics. Legislators may not have revisited the issue of the appropriate balance of power very often. When the issue was revisited, considerable weight may have been placed on how things always had been done.

The arrow in figure 1.1 runs from political competition in the state legislature to state courts, because state courts have historically had little influence on political competition in the legislature, whereas state legislatures have historically had tremendous influence on state courts. This influence arises because state legislatures make laws regarding the operation of state courts and provide funding for state courts.

Landes and Posner (1975) offer a theory linking levels of political competition to the structure of state courts. Their theory was formalized by Maskin and Tirole (2004) and Hanssen (2004b).[11] These authors argue that a dominant political party will prefer a more subordinate, less independent judiciary. For example, judges who face competitive reelection are likely to be more deferential to party officials in their decision making than the same judges would be if they held lifetime appointments. As political competition rises, however, the majority political party may at some point prefer appointed judges, because they may be more likely to preserve the majority party's policy legacy if the party becomes the minority party. The threshold will depend in part on the legislature's ability to screen for judges whose beliefs and preferences match those of the party in power. If they can perfectly screen for judges, then the

[11] See also Epstein et al. (2000) and Ramseyer (1994).

legislature will be indifferent between elections and appointment, because these judges will behave in the same way under elections and appointment. If screening is imperfect, however, then increases in political competition will at some point lead legislatures to switch to appointment. This model is extended to allow legislatures with different colonial legal systems to have different preferences regarding judicial independence. The primary implication is that the threshold level of political competition required to induce the state legislature to switch to a more independent judiciary is higher in civil-law states than in common-law states.

Using data on judicial retention systems, judicial tenure, the adoption of intermediate appellate courts, and judicial budgets, we investigate the implications of the model. Civil-law and common-law states differ in ways predicted by the model. Because it retained civil law, Louisiana is excluded from the analysis. Civil-law states had less independent state high court judges and lower expenditures on their courts. They adopted intermediate appellate courts—which provide more oversight of lower courts—at lower population levels. Civil-law states required larger increases in political competition to move away from partisan elections and to increase judicial independence more generally. Finally, they responded to changes in judicial independence differently. Common-law states tended to increase judicial budgets when they moved from election-based to appointment-based retention, while civil-law states tended to hold budgets constant. These differences hold even when controls are included for slavery and the timing of entry into the union. The results are similar if the years of civil law are used instead of a variable for whether the state had a civil-law or common-law colonial legal system.

The influence of civil law persisted for two related reasons. The model suggests that legislatures will not have incentives to make changes to judicial independence if levels of political competition are relatively stable. Moreover, in practice making a change to retention procedures does not only involve the legislature. Retention procedures for high court judges are specified in state constitutions. So any changes typically have to be ratified by the state constitutional convention or voters or both. These two factors imply that change will be comparatively rare.

We conclude that initial conditions played early and enduring roles in shaping political and legal institutions in the American states. Having made this argument, the book examines the effect of state political and legal institutions on an important economic outcome, state income per capita. Many of the initial conditions that influence political competition also influence income, so it desirable to try to separate the two effect. Per capita income in 1900 will capture, albeit imperfectly, the suitability of the state for agriculture and trade. If we control for per capita income in 1900, judicial independence has a strong positive association with per capita income in 2003. State political competition is not directly important for per capita income in 2003, in part, because it is hard to disentangle political competition and state per capita in 1900. However, a more through exploration of the effects of political competition and judicial independence on economic and social outcomes is beyond the scope of this book. The evidence from state per capita income adds further credence to the idea that political and legal institutions are important for long-run growth.

The book makes contributions to the literatures on institutions, American political history, and American legal history. It contributes to the literature on institutions by providing detailed evidence on the persistence of institutions over a long period of time in a large number of geographic units, on the mechanisms through which initial conditions shaped early institutions, and on the reasons for persistence. No work that we are aware of has been similar in scope. Previous research in this area, while extremely interesting, has provided limited information on persistence, mechanisms, and reasons for persistence. For example, Acemoglu, Johnson, and Robinson (2001) document that the initial conditions faced by European settlers in European colonies are strongly associated with the quality of institutions protecting property rights at the start and the end of the twentieth century. Banerjee and Iyer (2005) and Iyer (2010) highlight the relationship between British colonial institutions and measures of institutions and outputs in India.[12] Sokoloff and Engerman (2000) use proxies in their analysis of the role that initial conditions played in shaping institutions in North and South America. In all of these cases, the data are quite sparse for the period before the late twentieth century.

The book adds to the literature on American political history by offering a richer conception of what influenced the historical trajectories of state legislatures.

[12] They document the "institutional overhang" of British taxation systems and British direct and indirect rule that were eliminated in the mid-nineteenth century. This book documents and explains the persistent influence of defunct legal origins. While Banerjee and Iyer focus on the influence of history on economic outcomes, this book focuses on how history has influenced the evolution of political and legal institutions. Our book is also related to the work of Dell (2009), who analyzes the impact of history on districts within Peru. Dell documents that the mita, a forced labor system imposed by Spain on Peru and Bolivia between 1573 and 1812, has affected contemporary child stunting and consumption in matched (bordering) districts. She then argues, using available historical data, that the influence of the mita has persisted through its impact on land tenure and roads. Dell focuses on the impact of a particular historic institution for mobilizing labor on economic outcomes. Our study focuses on the impact of initial conditions on the evolution of broadly defined state institutions, including courts and legislatures.

The vast majority of the political history literature examines politics in a specific body, the U.S. Congress; in specific locations such as cities, counties, and states; or over specific, usually short, time periods.[13] To the extent that works are comparative and focused on multiple state legislatures over longer periods, much of the attention has been devoted to the split between the North and the South following the Civil War or to dating specific shifts in state politics.[14] This book goes beyond the North-South split and shows how the early occupational homogeneity of the elite also shaped the subsequent evolution of political competition in the states. Early occupational homogeneity helps explain variation in levels of political competition across states in the North and the South. The book also provides evidence that the persistent effect of these elites was the result of economic and political factors, including the malapportionment of geographically based political districts.

Finally, the book adds to the literature on American legal history by offering a more nuanced story of the development of American state courts that allows a greater role for colonial legal history. Scholars have written about the transition from civil law to common law in specific states and about the continuation of certain civil-law practices primarily relating to marriage and property.[15] But the dominant narrative has been quite dismissive of the effects of civil law, arguing that any effects were swept away by the tidal wave of Americans who entered the territory previously held by France, Spain, or Mexico.[16] This book suggests that the story is more complicated. Civil law appears to have influenced the balance of power between the state legislature and state courts in these states. The structure and funding of court systems in civil-law states are systemati-

[13] This literature is vast and interesting, but it is not about the historical trajectory of state legislatures.

[14] Despite Key's (1949) seminal work, remarkably little comparative work has been done on explaining differences in levels of political competition across states over long periods of time. Scholars have tended to focus on differences in competition as explaining differences in policy outcomes over a relatively short period. To the extent that differences in political competition are analyzed, the explanatory variables tend to be contemporary demographic variables. For example, Patterson and Caldeira (1984) take up precisely the question that we are interested in, but examine a short period of time and use demographic variables to explain differences. Elazar (1966) takes up this question as well and considers historical factors. Unfortunately, his work is largely descriptive in terms of political culture, its relationship to political competition, and its relationship to political outcomes more broadly. As is discussed in chapter 3, his political culture variable is almost linearly related to average state temperature.

[15] See Arnold (1985), Banner (2000), Cutter (1995), Fernandez (2001), Langum (1987), Bakken (2000), and Friedman (1986).

[16] See Friedman (1986). Other major legal histories of the period such as Horwitz (1977) never even discuss civil law.

cally different than in the court systems in common-law states. The book provides evidence that civil-law state legislatures made changes to their judicial retention systems under different conditions than common-law state legislatures did. Moreover, the response in terms of expenditures following changes in retention systems differed across common-law and civil-law states.

The book begins by describing the initial conditions, then discusses states legislatures, state courts, and finally per capita income. Colonial legal systems are the subject of chapter 2, and chapter 3 introduces the other initial conditions. This ordering allows us to investigate whether colonial legal systems were related to political competition in the state legislature. Chapter 4 presents the mechanism, occupational homogeneity of the elite, through which initial conditions acted on political competition in the state legislature. Chapters 5 and 6 present a theory of how political competition and colonial legal systems influence judicial independence, and they investigate empirically the extent to which the two factors influenced judicial independence. Chapter 7 links political and legal institutions to state per capita income.

CHAPTER 2

Legal Initial Conditions

THIS CHAPTER EXAMINES colonial legal systems and their effects on the balance of power between the state legislature and the state high court. Why is the balance of power important? It is important because the balance of power determines the extent to which the state high court can act as a constraint on the legislature and the elites more broadly. Establishing and maintaining an appropriate balance of power has been and remains a critical issue at both the state and the federal levels.

Civil law and common law had and still have different visions of the appropriate balance of power between the legislature and the courts. Civil law views the courts as subordinate to the legislature, while in common law the two institutions are more equal. This statement is relatively uncontroversial when comparing, for example, French civil law with British common law after the French Revolution. In this chapter we evaluate the extent to which these civil-law and common-law visions of the balance of power differed in United States in the eighteenth and nineteenth centuries. The available evidence suggests that the visions did differ, and that civil law viewed judges and courts as subordinate to the legislature.

To understand the effect of colonial legal systems on the balance of power between the state legislature and the state high court, one first needs to determine what the legal initial conditions were. We conduct a detailed examination of the settlement history and historical operation of courts in what would become the American states. The result is that we classify thirteen states as having civil-law colonial legal systems. Twelve of these states would eventually adopt common law. The thirteenth, Louisiana, retained civil law.

Unfortunately, the balance of power between the legislature and the judiciary cannot be observed directly. For these thirteen civil-law states, there is evidence that individuals with civil-law backgrounds were in the right place at the right time and so plausibly could have influenced the balance of power. They may not have been the only actors. Individuals with common-law backgrounds may have preferred a weaker judiciary as well. In states with common-law histories, their ability to act on this preference may have been limited. A state's civil-law background may have offered an opportunity for them to act on this preference.

Chapters 5 and 6 document that courts in civil-law and common-law states differed along a number of dimensions over the period 1866–2000 and that these differences were consistent, with common-law states having more independent judiciaries than civil-law states. These differences are associated with differences in judicial decision-making. Chapter 7 presents evidence linking judicial independence to state income to per capita.

Balance of Power

Strong courts are generally thought to be good for economic activity and society more broadly. The reason is that they are likely to be more successful than weak courts at constraining the behavior of legislatures and elites. In the absence of constraints, legislatures and elites may engage in activities that are good for them individually or collectively but are not good for the community as a whole. Thus, differences in the balance of power may have economic and social implications.

Concerns about balance of power date to the earliest days of the country. William Nelson (2008) describes the evolution of the balance of power in Massachusetts before 1660: "As early as 1635, the townspeople of Massachusetts . . . demanded that their law be reduced to a detailed written code that would be accessible to them. This movement for codification resulted from two separate though overlapping urges—first, the colonists' 'distrust of discretionary justice' as administered by their betters, and second, the 'importance' they 'attached to stable and written laws' that they could readily ascertain and obey."[1] In 1649 after a series of unsuccessful attempts to balance the power of the townspeople and the elite in the General Court, which despite its name was both a legislative and a judicial body, representatives of the communities were given the final word on judicial disputes. This power acted as a constraint on the magistrates and through the magistrates on the elites more broadly. Nelson concludes: "In Massachusetts, the purpose of the rule of law . . . was to control the discretion of the governing elite and thereby prevent it from interfering unpredictably in the lives of ordinary people in the local towns."[2]

Massachusetts was not unique—every state had to struggle with issues of balance of power. Some states were more successful than others. And within states, the constraints were more successful in some periods than other. Part of the issue was that the nature of the constraints changed over time. Early

[1] Nelson (2008), 72.
[2] Nelson (2008), 78.

juries in Massachusetts and other colonies had far more power than they have today. But as law became more complex, the role of the jury changed, and other means had to be developed to strengthen the judiciary. The struggle to constrain the elites was ongoing process.

The federal government also had to struggle with balance of power. The importance of balance of power was recognized by the founding fathers, who among other things gave federal judges life tenure. Although it is easy to take the power of the Supreme Court for granted today, its power was not a foregone conclusion. The early Court owes a tremendous debt to John Marshall, who was able to reach a decision in *Marbury v. Madison* (1803) that simultaneously strengthened the Supreme Court and averted an ugly political battle.[3] Because of his status, ambition, and political acumen, the Court grew stronger during his lengthy tenure as chief justice (1801–1835). Even with this start, the Supreme Court generally acted as less of a constraint on Congress and the president during the nineteenth and early twentieth centuries than it does today. The Court survived a major conflict with President Roosevelt over the unconstitutionality of New Deal legislation. Under the threat of expansion under Roosevelt's Court-packing plan, the Court began to rule that New Deal legislation was constitutional. The power of the Court generally grew during the remainder of the twentieth century.

Colonial Legal Systems and Balance of Power

How were colonial legal systems likely to shape the balance of power? The two dominant legal traditions, civil law, which is used by most European countries and their former colonies, and common law, which is used in Great Britain and its former colonies, are widely viewed as differing along a number of dimensions.[4] The most salient of these differences for our study is that from at least the late eighteenth century on, civil-law legal systems have had relatively more powerful legislatures and more subordinate judiciaries than common-law legal systems.[5] If the colonial legal system had an impact on the early balance of

[3]For more detail on *Marbury v. Madison* and its context, see Nelson (2000).

[4]See Merryman and Perez-Perdomo (2007) and Zweigert and Koetz (1998). The relevant point to draw from their book and the very large literature debating the convergence or lack of convergence between the two legal systems in the late twentieth and early twenty-first century is that virtually all scholars appear to agree that the two systems were very different in the nineteenth century and for much of the twentieth century.

[5]Although common-law judges in England were technically subordinate to Parliament, in practice they were independent. Historically Parliament was widely recognized as having the power to

power between the legislature and the judiciary, it should cause civil-law states to have a weaker judiciary than common-law states.

For differences in legal systems to have led to differences in the balance of power, the two legal systems had to have differed philosophically and operationally in the United States during the relevant time period. The original British colonies and their successor states were subject to English common law and later American common law.[6] The relationship between these legislatures and their courts likely reflected the views of the period. That is, legislatures generally respected courts and judges and afforded them considerable independence. Moreover, the legislature accepted this as the "correct" relationship. In states that entered the union later, many of the original settlers were American or English by birth.[7] These settlers and subsequent settlers transferred their views on the appropriate balance of power between legislatures and courts to these new locations.

The literature on civil-law legal systems as they operated in the late eighteenth and early nineteenth centuries in the United States is sparse. However, the existing literature allows some conclusions to be drawn.

In principle, written law in the form of colonial codes governed both Spanish and French colonies. The most prominent of these for Spanish administrators was the *Recopilación de Indias* (1680), which was specifically designed for the administration of Spanish possessions in the New World. In fact, relatively few copies of the codes appear to have existed in the colonies.[8] The written orders of commandants and higher officials provided an additional source of law. One of the most important of these was the Code O'Reilly (1769). It codified written

---

write and revise statutes. It is worth noting, however, that judges often played an influential role in writing statutes. Further, judges had considerable latitude in interpreting statutes and in setting and revising precedent in the absence of statute. In the early seventeenth century, the question of judges' ability to declare laws void was debated, but Parliament remained supreme in this sense. See Baker (2007), 208–212.

The structure of the judiciary promoted judicial independence. During the seventeenth and eighteenth centuries, judges were allowed to compete for cases, setting their own fees. From the end of the eighteenth century, judges had salaries, but also life tenure, which gave them greater independence. See Klerman (2007) on judicial competition and Klerman and Mahoney (2005) on judicial independence. Through the mid-nineteenth century, judges used legal fiction as a strategic tool to prevent appellate review, which reinforced their de facto independence. See Klerman (2009) on legal fiction.

[6]See Friedman (1986), Horwitz (1977), Hoffer (1992).

[7]A note on terminology is in order. References to certain locations being unpopulated or first settled by should be read as being unpopulated by or first settled by Europeans. In nearly all cases Native Americans were present, sometimes in high density.

[8]There appears to be some debate on this point, with both Banner (2000) and Langum (1987) arguing for a scarcity of texts in Missouri and California, respectively, and Cutter (1995) arguing for their existence in Texas and New Mexico.

law for what would later become Alabama, Arkansas, Louisiana, Mississippi, and Missouri and remained in force until 1803.[9]

In practice much of the law appears to have consisted of unwritten norms.[10] To the extent that it is possible to tell, French and Spanish civil-law norms appear not to have been very different in practice.[11] This is relevant because Spain governed most of the land encompassed by the Louisiana Purchase from 1762, when it acquired the region from France, until 1803, when the French briefly regained control, having officially reacquired the territory in 1800. Spanish civil law also heavily influenced other states. Following early control by the French, Florida was governed by Great Britain and then Spain, and Arizona, California, New Mexico, and Texas were governed by Spain and then Mexico.

Whether written or unwritten, colonial civil law was distinct from common law. At the highest level, colonial judges were usually military or civilian governors or senior administrators. French, Spanish, and Mexican administrators were part of a large bureaucracy for administering colonial possessions and thus typically had considerable military, administrative, and judicial experience from previous postings. These administrators acted as judges in ways consistent with civil law of the period. Their role was inquisitorial. In more significant proceedings, petitioners submitted written pleadings.[12]

Furthermore, these officials appear to have conceptualized the job as largely administrative in the sense that it involved applying existing civil-law precepts. Although they date from the mid-eighteenth century, it is useful to quote from the instructions given to senior Spanish officials (*corregidores* and *alcaldes mayores*) bound for the colonies:

> You will keep and execute, precisely and exactly, all the orders, *cédulas*, compilations of laws, and all that which is ordered and might be ordered for the proper governance of this office, the good treatment of the Indians, and the better administration of justice. Thus, you must hear all lawsuits and cases—both civil and criminal—that might occur in your jurisdiction, which you can and must hear, and order all things that your predecessors might have ordered and provided. And also you will receive all judicial inquiries and information in the cases and things permitted by law that

[9]France took nominal control in 1800, but did not take actual control until 1803.

[10]See especially Banner (2000), 37, and chapter 3. The discussions in Langum (1987) and Cutter (1995) also discuss norms at some length. The use of norms was specifically allowed for in civil law during this period.

[11]Arnold (1985), 44, and Fernandez (2001), 15.

[12]See Arnold (1985), Banner (2000), Cutter (1995), Fernandez (2001), Langum (1987).

you believe might be conductive to my royal service, good governance, and the administration of justice.[13]

These officials were there to apply royal law and when necessary to make laws useful for promoting good governance.

Below these bureaucrats were other individuals who acted as judges, since high-level bureaucrats simply did not have the capacity to hear all of the cases. These lower level judges tended to be well-respected members of the community. While typically lacking in formal legal training, they were nonetheless able to draw on common sense and local practice to handle the vast majority of ordinary cases. Since many held their positions either formally or informally at the pleasure of the governor, they could be removed for poor conduct. Thus they too fell under bureaucratic control.

One critique of the view that civil law could have had a persistent influence on American state legal systems relates to timing. This issue arises because a few states that will be classified as civil-law states were acquired by Great Britain or the United States prior to the French Revolution in 1789, and others were acquired fairly shortly thereafter. The history of these states and the reasons for classification will be discussed in detail shortly.

In particular, there is a debate over when judges became relatively subordinate in civil law.[14] Dawson (1960) and more recently Glaeser and Shleifer (2002) argue that differences in the role of the judiciary in common-law and civil-law legal systems originated during the twelfth and thirteenth centuries. Dawson (1960) demonstrates that the judiciary in France was already much more centralized and bureaucratic than England well before the Revolution. Klerman and Mahoney (2007) have argued that the differences originated later, predominantly around the time of the French Revolution. It is worth noting that the effect of the Revolution was immediate. Legislation subordinating civil-law judges was passed in 1790, and their status was reinforced by the Constitution of 1791.[15] The publication of the Napoleonic Code in 1804 ratified the status quo.[16]

The debate regarding the timing of the subordination of judges in France sheds little light on actual levels of independence of civil-law and common-law judges in what would become the United States. Because Spain governed ten of the thirteen civil-law states between the 1760s and the early nineteenth century,

[13] Quoted in Cutter (1995), 76.

[14] See Berman (1983), Glaeser and Shleifer (2002), Klerman and Mahoney (2007).

[15] Stone Sweet (1992), 25.

[16] The more heavily populated parts of these states became American in 1798. In this case, they would have been acquired before the publication of the Napoleonic Code.

it is useful to consider the Spanish colonial experience. In the colonies, Spanish military or civilian administrators exercised both legislative and judicial power. Importantly, around 1810 all former Spanish colonies in Latin America developed systems of governance in which all authority rested in the legislature.[17] While this does not directly bear on the civil-law states, these systems of governance are indicative of colonial residents' conceptions of the appropriate level of judicial independence. It is possible that these former Spanish colonies were heavily—or even dramatically—influenced by the publication of the Napoleonic Code. If this is true, their experience is not relevant for understanding the experience of the eight civil-law states formed from territory that had been acquired up to 1804. Our suspicion is that the groundwork for these differences was laid much earlier.

Thus, for timing to be problematic, one would need to believe two things. First, in the late eighteenth century in what would become the United States, civil-law judges had a level of judicial independence similar to or greater than their common-law counterparts. And, second, changes in civil-law attitudes about the appropriate relationship between the legislature and the judiciary had an impact on North America only after publication of the Napoleonic Code in 1804. Explicit comparison of civil-law and common-law judges' independence in North America after 1776 is difficult, because civil-law judges were colonial and common-law judges were not. For that reason alone, common-law judges would be more independent. However, by 1803, in addition to preexisting civil-law views on the independence of the judiciary, the precepts of the French Revolution had had more than a decade to filter into Spanish colonial administration through various channels and into North America directly.[18] Thus it seems unlikely at the time of acquisition that residents in civil-law areas held views on the appropriate level of judicial independence from the legislature that were similar to those of individuals in common-law areas.

Effects on State Courts

How are differences in balance of power likely to be manifest in state courts? To the extent that colonial legal systems affected the early balance of power and the effect persisted, states with civil-law legal systems should have less

[17] Dealy (1968), 46ff. discusses the lack of a conception of balance of power in Spanish colonial law and in former Spanish colonies. All of the power was lodged in a single entity—the legislature.

[18] Cutter (1995, 23) notes that the pressures of enlightened despotism and increased regal control had filtered into the Spanish colonies.

independent judges, may have smaller budgets, and will likely have different legal outcomes than states with common-law legal systems. Chapters 5 and 6 present evidence consistent with common-law and civil-law states having had different balances of power.

Classification

From the time of Christopher Columbus's arrival in North America in 1492, European powers vied for footholds. On the eastern seaboard, territorial rivals included England, France, the Netherlands, Spain, and Sweden. Of the five rivals, four had civil-law legal systems. As a result of the ongoing struggle, individuals from a variety of backgrounds were constantly moving about, particularly on the frontier. In addition, there was sometimes a foreign military presence.

Figures 2.1 and 2.2 show the regions of North American that various nations controlled at the end of the seventeenth and eighteenth centuries. The territories had shifted quite significantly over this one-hundred-year span. They would shift markedly again during the first half of the nineteenth century. Figure 2.3 shows the territory occupied by the forty-eight continental states and the dates of territorial acquisition. With the exception of the Oregon Country, which was British, all of the territory west of the Mississippi was acquired from civil-law countries. In addition, Florida and parts of Mississippi and Alabama were also acquired from civil-law countries. Although it is not obvious from figure 2.3, considerable territory east of the Mississippi was acquired by Britain from France during the eighteenth century.

In classifying states, the historical record was examined to determine which country first permanently settled a state and whether there were operational courts in the state. Classification is fairly clear-cut for a state such as Massachusetts. It was first permanently settled by the British, was under British control up to statehood, and had operational British colonial courts. Thus, it was always a common-law state. Some states, such as Nebraska and Nevada, were first permanently settled by the United States and never had civil-law courts. They are classified as common law. Similarly, classification of Louisiana and California as civil-law states is straightforward because both had substantial permanent populations and operational civil-law courts at the time of U.S. acquisition. Other states require some judgment in classification, because very early permanent or impermanent settlement by civil-law countries was followed by British or American settlement.

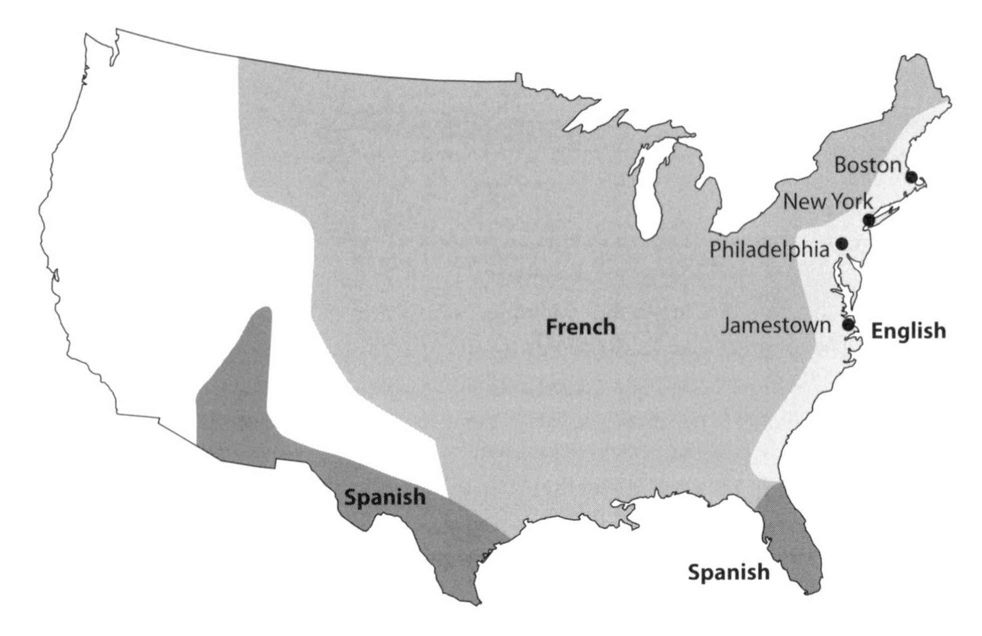

Figure 2.1. European Colonies at the End of the Seventeenth Century.

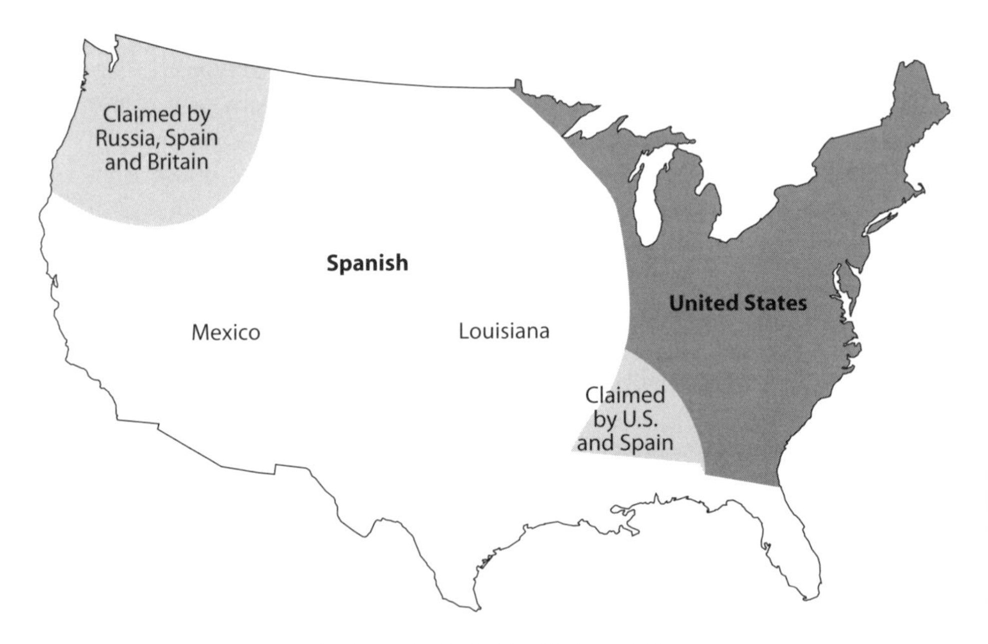

Figure 2.2. European Colonies at the End of the Eighteenth Century.

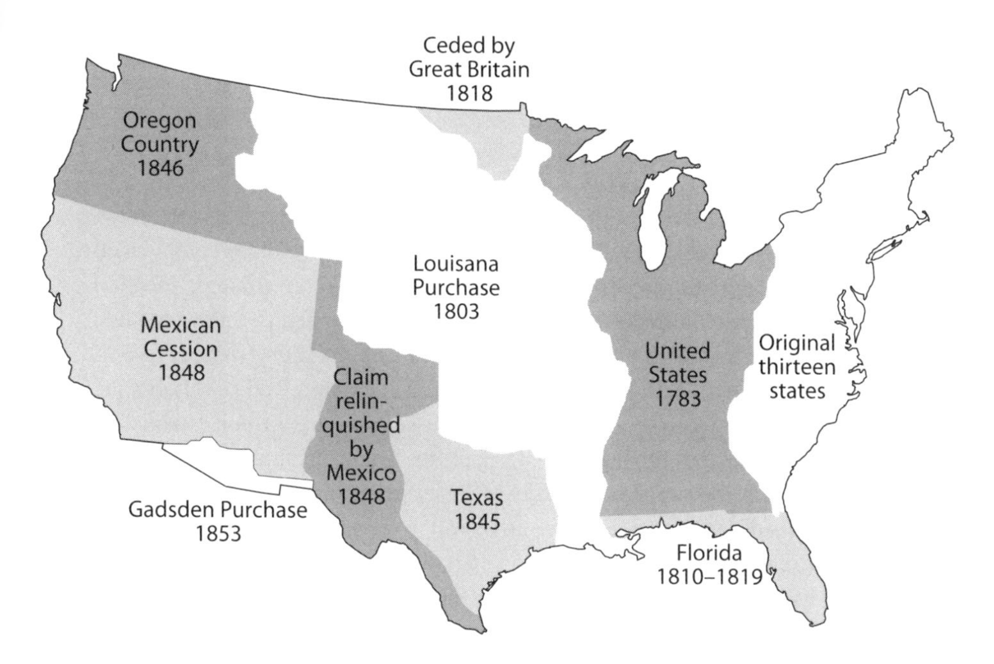

Figure 2.3. American Territorial Acquisition.

New England

England permanently settled and maintained control over the area covering the six New England states—Connecticut, Massachusetts, Maine, New Hampshire, Rhode Island, and Vermont.[19] The French were frequently present in the northern border states, notably Maine, New Hampshire, and Vermont, but their presence never seems to have gone beyond military control at isolated forts. In particular, unlike the Mississippi Valley and other regions, there was no evidence of substantial settlement or of operational courts. This is not to say that settlers in those northern states and even settlers in Massachusetts, Connecticut, and Rhode Island were not concerned about French influence. French incursions from Montreal and relations with the Indians posed a serious and ongoing threat for much of the seventeenth and eighteenth centuries. But permanent French settlements

[19]This discussion is based on Dwight (1842), Trumbull (1898), and Clark (1914) for Connecticut; Maine Historical Society (1919) and Abbott (1892) for Maine; Austin (1875) and Bradford (1835) for Massachusetts; Belknap and Farmer (1862) and Barstow (1842) for New Hampshire; Greene (1877) and Taylor (1853) for Rhode Island; Collins (1903) and Hall (1868) for Vermont.

and operational civil-law courts never emerged. Thus in table 2.1 all six states are classified as common law.

The Mid-Atlantic

Compared to the New England states, the four states in the mid-Atlantic—Delaware, New Jersey, New York, and Pennsylvania—have more complicated histories.[20] Dutch and Swedish settlements occupied a key strategic location in the mid-Atlantic during the seventeenth century. The first permanent Dutch settlements were established in 1624 in Albany and on High Island (present-day Burlington Island) in the Delaware River. Permanent Swedish settlements followed in 1638, when settlers under the command of Peter Minuit established Fort Christina. Within the next two decades, the Dutch established permanent settlements in Delaware, New York, New Jersey, and Pennsylvania. The Swedish established permanent settlements in Delaware and Pennsylvania.[21] Both the Swedish and Dutch settlements had operational civil-law courts.[22] In 1655, the Swedish settlements were captured by the Dutch and became part of Dutch New Netherland. In 1664, the British captured New York. That year the population of Dutch New Netherland was estimated to be 9,000.[23]

Although it was briefly recaptured by the Dutch, New Netherland was basically under British control from 1664 onward. The terms of acquisition were initially favorable to the Dutch settlers, largely preserving the existing Dutch

[20]This discussion is based on Scharf (1888) and Vincent (1870) for Delaware; Gordon (1834) and Smith (1877) for New Jersey; Morgan (1851) and O'Callaghan (1848) for New York; Gordon (1829) and Sharpless (1900) for Pennsylvania.

[21]The Dutch also had a settlement in Connecticut that appears to have been temporary. Estimates of the fraction of the population with Dutch surnames in the 1790 census indicate that by far the largest or at least the most enduring populations were in New York and New Jersey (Purvis 1984).

[22]For court records pertaining to New York and New Jersey, see Van Laer (1974) for court records of the Director General and Council of New Netherland, 1638–1664 (the highest court, covering all of New Netherland); and Fernow (1897) for court records of the Courts of Schouts and Schepens for New Amsterdam. For court records pertaining to Pennsylvania and Delaware, see Brodhead and O'Callaghan (1853), volume 12 for Dutch minutes of court actions, 1655–1657; Armstrong (1969) for records of the Upland Court (Chester County, Pennsylvania), 1676–1681, which was a Dutch court that continued to operate after English acquisition; Gehring (1981) for the Dutch records, 1648–1664; and Johnson (1930) for some court minutes for New Sweden, 1643–1644.

[23]See Rink (1986). In 1673, the Dutch temporarily regained control of New Netherland. The land was officially ceded to England in 1674.

legal system. In 1665, however, the Dukes Laws, modeled on New England legal codes, were imposed on the New York colony. This consolidated British control of a large stretch of the Atlantic seaboard. Under the Treaty of Westminster, the British acquired permanent control of Dutch New Netherland in 1674.

As in the more northern states, the French were of considerable concern during the eighteenth century. The French had many Native American allies, who resented British encroachment. Much of the activity was concentrated in northern and western New York and in western Pennsylvania. Their capture of Fort Duquesne (located in present-day Pittsburgh) only heightened the concern. None of the incursions, however, resulted in permanent French settlement and operational French courts.

Given that all four mid-Atlantic states meet our criteria for being classified as civil law, why are they listed as common law in table 2.1? Civil law was abolished very early, certainly by the late seventeenth century. This alone raises questions because of the debate regarding the timing of subordination of judges in civil law. Suppose, however, that Dawson (1960) and Glaeser and Shleifer (2002) are correct and that the two systems had diverged in their treatment of judges by the 1660s. Further, suppose that Dutch civil law in its colonies was closely related to French civil law.

Even if the two systems differed, the transmission of civil-law attitudes remains problematic. If Dutch settlements shaped the early balance of power between the judiciary and legislature, and this relationship persisted, then Dutch civil-law views of the appropriate balance of power would need to have had a significant impact on the attitudes of the (British) colonial legislature. The persistence of a Dutch civil-law balance of power would require that these attitudes endured and affected the attitudes of the newly formed state legislature a hundred years later. Alternatively, and even less plausibly, more than two generations of individuals with civil-law heritage would have had to grow up in a culture that specifically identified British judges as "too powerful" and reined in this power in the constitutional convention or the early legislature. By the time of American independence, these four states had had more than one hundred years of experience under British colonial rule.

The Upper South

The eight states that fall loosely into the Upper South—Georgia, Kentucky, North Carolina, South Carolina, Tennessee, Virginia, and West Virginia—are straightforward to classify, because the only settlement by civil-law countries

Table 2.1.
Classification of American States by Colonial Legal System

State	*Permanent settlement by civil-law country*	*Operational civil-law courts*	*Classification*
New England			
Connecticut	N	N	Common law
Maine	N	N	Common law
Massachusetts	N	N	Common law
New Hampshire	N	N	Common law
Rhode Island	N	N	Common law
Vermont	N	N	Common law
The Mid-Atlantic			
Delaware	Y	Y	Common law
New Jersey	Y	Y	Common law
New York	Y	Y	Common law
Pennsylvania	Y	Y	Common law
The Upper South			
Georgia	N	N	Common law
Kentucky	N	N	Common law
Maryland	N	N	Common law
North Carolina	N	N	Common law
South Carolina	N	N	Common law
Tennessee	N	N	Common law
Virginia	N	N	Common law
West Virginia	N	N	Common law

TABLE 2.1. *(continued)*

State	*Permanent settlement by civil-law country*	*Operational civil-law courts*	*Classification*
The Midwest			
Illinois	Y	Y	Civil law
Indiana	Y	Y	Civil law
Iowa	N	N	Common law
Michigan	Y	Y	Civil law
Minnesota	N	N	Common law
Missouri	Y	Y	Civil law
Ohio	Y	N	Common law
Wisconsin	Y	N	Common law
The Lower South			
Alabama	Y	Y	Civil law
Arkansas	Y	Y	Civil law
Florida	Y	Y	Civil law
Louisiana	Y	Y	Civil law
Mississippi	Y	Y	Civil law
The Unpopulated West			
Colorado	N	N	Common law
Idaho	N	N	Common law
Kansas	N	N	Common law
Montana	N	N	Common law
Nebraska	N	N	Common law

TABLE 2.1. *(continued)*

State	*Permanent settlement by civil-law country*	*Operational civil-law courts*	*Classification*
Nevada	N	N	Common law
North Dakota	N	N	Common law
Oklahoma	N	N	Common law
South Dakota	N	N	Common law
Utah	N	N	Common law
Wyoming	N	N	Common law
The Populated West			
Arizona	Y	Y	Civil law
California	Y	Y	Civil law
New Mexico	Y	Y	Civil law
Oregon	N	N	Common law
Texas	Y	Y	Civil law
Washington	N	N	Common law

Note: For discussion of the classification of each state, see text.

was early and very short-lived.[24] In 1562, the French had established a small settlement in South Carolina. Within a few years, the Spanish captured the settlement, killed the French settlers, and resettled it. Shortly thereafter, the French returned, destroyed the Spanish settlement, killed the inhabitants, and returned to France.[25] Permanent settlement would come later under the English. Because France and Spain never permanently settled in these states or operated civil-law courts, states in the Upper South are all classified as common law.

The Midwest

The eight states in the Midwest—Illinois, Indiana, Iowa, Michigan, Minnesota, Missouri, Ohio, and Wisconsin—had considerable contact with the French as the result of French control of the Mississippi Valley and their presence around the Great Lakes.[26] Unlike many of the border regions in New England and the mid-Atlantic, there were some permanent French settlements.

Evidence on the existence of these settlements comes from a number of sources, including primary documents and governmental records. Unfortunately, these sources generally do not provide estimates of population for these French settlements. Some indirect evidence on population has, however, survived. Table 2.2 lists the number of foreign land grants that were confirmed and patented. Because of fraud and changing standards for confirmation and patenting, the number of confirmed foreign land claims is at best a noisy measure of the number of adult male settlers. Even allowing for these factors, there were a substantial number of land grants in four states—Illinois, Indiana, Michigan, and Missouri. The remaining states had few or no land grants.

Table 2.3 provides estimates of postacquisition, and where available preacquisition, population for the four states with larger numbers of land grants. The

[24]This discussion is based on Arthur and Carpenter (1853) and Jones (1883) for Georgia; Kinkead (1896) and Arthur and Carpenter (1852) for Kentucky; Bozman (1837) and McSherry and James (1904) for Maryland; Ashe (1908) and Moore (1880) for North Carolina; Ramsay (1809) and McCrady (1897) for South Carolina; Phelan (1888) and Haywood, Colvar, and Armstrong (1891) for Tennessee; Beverley and Campbell (1722) and Stith (1865) for Virginia; Lewis (1887) and Callahan (1913) for West Virginia.

[25]Spain and France engaged in armed conflict in South Carolina and Florida in 1565, as part of the Spanish plan to regain control of the North Atlantic coast. Vigneras (1969).

[26]This discussion is based on Reynolds (1887), Wallace (1893), and Alvord (1920) for Illinois; Esarey (1915) and Dillon (1859) for Indiana; Sage (1974) and Cole (1921) for Iowa; Burton, Stocking, and Miller (1922), Parkins (1918), and Lanman (1841) for Michigan; Neill (1858) and Folwell (1921) for Minnesota; Houck (1908) and Violette (1918) for Missouri; Atwater (1838) and Abbot (1875) for Ohio; and Tuttle (1875) and Smith (1854) for Wisconsin.

Table 2.2.
Confirmed Private land Claims in American States as of June 30, 1904

State	Number of claims	Area of claims in acres	Average claim size in acres
New England, Mid Atlantic, and Upper South Not part of the public lands system			
The Midwest			
Illinois	936	185,774	198
Indiana	862	188,304	218
Iowa	1	5,760	5,760
Michigan	942	280,673	298
Minnesota	0	0	0
Missouri	3,748	1,130,052	302
Ohio	111	51,161	461
Wisconsin	175	32,779	187
The Lower South			
Alabama	448	251,602	562
Arkansas	248	110,090	444
Florida	869	2,711,291	3,120
Louisiana	9,302	4,347,891	467
Mississippi	1,154	773,087	670
The Unpopulated West			
Colorado	6	1,397,886	232,981
Idaho	0	0	0
Kansas	0	0	0
Montana	0	0	0

TABLE 2.2. (*continued*)

State	Number of claims	Area of claims in acres	Average claim size in acres
New England, Mid Atlantic, and Upper South Not part of the public lands system			
Nebraska	0	0	0
Nevada	0	0	0
North Dakota	0	0	0
Oklahoma	0	0	0
South Dakota	0	0	0
Utah	0	0	0
Wyoming	0	0	0
The Populated West			
Arizona	95	295,212	3,107
California	588	8,850,144	15,051
New Mexico	504	9,899,022	19,641
Oregon	7,432	2,614,082	352
Texas	Handled by the Texas state government		
Washington	1,011	306,796	303

Source: From the *Report of the Public Lands Commission*, http://memory.loc.gov/gc/amrvg/vg57old/vg57.html, image 84.

Notes: Utah (60 grants totaling 8,876.80 acres) was excluded because no documentary evidence could be found indicating the source of these land claims. In particular, there was no evidence to suggest that they were confirmed as part of the work of the Surveyor General of the New Mexico Territory or the Court of Private Land Claims, which were responsible for addressing claims in all territory acquired from Mexico other than California. Land grants in Oregon and Washington were donations made to compensate existing settlers for land they held under Britain. Land grants for Texas are not reported because Texas was briefly independent. The strategy there was quite similar, however. Residents were given land, with earlier settlers receiving bigger parcels.

Table 2.3.
Estimates of Pre- and Postacquisition Population in Civil-Law States

State	*Year acquired*	*Date estimate*	*Population*	*First U.S. census*	*Pop. at first census*
The Midwest					
Illinois	1763	1763	<2,000	1800	2,458
Indiana	1763			1800	2,632
Michigan	1763			1800	3,757
Missouri	1803	1804	9,373	1810	19,783
The Lower South					
Alabama	1798/1813	1812	<1,000	1800[a]	1,250
Arkansas	1803	1798	400	1810	1,062
Florida	1821	1795	8,363	1830	34,730
Louisiana	1803/1810	1803	43,000	1810	76,556
Mississippi	1798/1813			1800[a]	7,600
The Populated West					
Arizona	1848/1853	1846	<1,000	1860	6,482
California	1848	1846	10,000	1850	92,597
New Mexico	1848/1853	1846	65,000	1850	61,547
Texas	1846/1848	1836	40,000	1850	212,592

Notes: Arkansas, Louisiana, and Missouri were acquired as part of the Louisiana Purchase. The northern portions of Alabama and Mississippi were part of the original territory acquired from Great Britain. The U.S. established control of western Louisiana and southern Alabama and Mississippi in 1810, 1812, and 1812. This territorywas formally acquired along with Florida from Spain in 1821. Parts of Arizona and New Mexico, all of California, and the questionable title to Texas, which had been independent and then opted to join the U.S. in 1846, were acquired from Mexico in 1848. Additional territory in southern Arizona and New Mexico was acquired as part of the Gadsden Purchase in 1853. The population estimates for Alabama are for the part of Alabama controlled by the Spanish. The largest city in this area was Mobile. See Hamilton (1910), 405 (for 1812), 447 (for 1818). The population estimates for Arizona, New Mexico, and Texas are from Weber (1982), 183–184 (Arizona), 195 (New Mexico), and 177 (Texas). The estimate for Arkansas is from Arnold (1985), appendix IV, p. 222. The estimate for California is from Langum (1987), 23, table 1. The estimate for Louisiana is from Dargo (1975), 6. The estimate for Missouri is from Banner (2000), 14 n. 9. The estimate for Florida is from Coker (1999).

[a]Above 31 degrees.

number of land grants is roughly consistent with the pre and postacquisition estimates. Taken together with other surviving records, these numbers suggest that there were substantial permanent settlements in these states. In Michigan, settlement occurred at Detroit on the Detroit River. In Indiana, settlers were located in Vincennes on the Wabash River. In Illinois, settlement was at Cahokia and Kaskaskia on the Mississippi River. The seat of French government for the Illinois Country was at Fort Chartres, near Kaskaskia. In Missouri, settlers were located at St. Louis and Ste. Genevieve on the Mississippi River. Other smaller settlements existed as well.

Historical evidence suggests that civil law was in force in Illinois and Indiana through 1789 and possibly longer. Records from the village assemblies, which governed many aspects of village life, and records of disputes that made it to New Orleans, which was the administrative center for France and later Spain for colonial holdings along the Mississippi, suggest that a formal judicial system operated in Illinois and Indiana.[27] Civil-law courts were operational in Kaskaskia and New Chartres in the 1750s.[28]

Britain nominally acquired the region from France in 1763, but it was slow to take control and generally unsuccessful at imposing common law on the French residents. A common-law court was established in Kaskaskia in 1768. The court was unpopular and appears to have blended common law and civil law to appease resentment. By 1774, civil law had officially returned as a result of the Quebec Act. The act gave residents of the province of Quebec, which included Illinois, the right to use civil law, although British common law was used for criminal cases.

In 1779 the Virginia government, probably because of its tenuous hold on the region, chose to continue the use of French civil law. George Rogers Clark captured Kaskaskia from the British in July 1778 and then famously recaptured it in February 1779. Virginia used this victory as the basis for its claim to Kaskaskia, Cahokia, Vincennes, and the associated territory. John Todd, who organized the government for Virginia in 1779, found it expedient to permit the continuation of the status quo, namely French civil law. Once Virginia withdrew, there was a power vacuum.

In the late 1780s, there were no official courts in operation, but in practice civil law seems to have prevailed.[29] In 1787, the Northwest Ordinance specifically

[27]On French Illinois, see Ekberg (1998) and Briggs (1990). Unfortunately, there was only rarely a notary in the Illinois Country, and what notarial records there may have been have not survived.

[28]Wallace (1893), 310. Alvord (1920) discusses the discovery and content of the surviving records.

[29]Alvord (1909), 383–384 and Alvord (1920), 336–338.

protected the property and inheritance rights of the French residents in Kaskaskia and Vincennes, suggesting both that they had some political power and that civil law was still in operation.[30] Certainly as late as 1789, Major Hamtramck, the head of the post at Vincennes in Indiana, wrote to General Harmar complaining about the lack of government, and specifically about the lack of courts.[31]

In Michigan, civil law prevailed until 1792. Unlike Illinois and Indiana, which were ruled from New Orleans, Michigan was ruled from Montreal or Quebec City. By 1730, Detroit had a royal notary, Robert Navarre.[32] Together with the commandant, he handled much of the routine legal business. Larger and more complex cases were forwarded on to Montreal or Quebec City. The British assumed control of Detroit in 1760. Much of the government was military, and most of the population was French. As a result, the British tended to use locals as notaries and judges. Navarre was asked to continue as notary, and later that decade Philip Dejean, a Frenchman, was appointed as a judge. Any use of common law—if there had been any—was relatively short lived. The Quebec Act (1774) covered Michigan, and so reinstated French civil law. French law would continue in force until 1792, when it was formally replaced by British common law. American forces took control of Detroit four years later.

Missouri requires less discussion regarding classification, because it had been under continuous control by the French and Spanish up to the time of American acquisition. Stuart Banner's book discusses in detail the operation of its civil-law courts and the transition from civil law to common law.[33]

In sum, the available historical evidence on the timing of the change in legal systems supports classifying Illinois, Indiana, Michigan, and Missouri as civil-law states.

What of the other four Midwestern states? A complete lack of evidence of permanent settlement by foreign governments in Iowa and Minnesota leads us to classify them as common law. Ohio and Wisconsin are interesting cases, because of the small number of land grants. Wisconsin was settled later and

[30]Section 2 of the Northwest Ordinance: "Saving, however to the French and Canadian inhabitants, and other settlers of the Kaskaskies, St. Vincents and the neighboring villages who have heretofore professed themselves citizens of Virginia, their laws and customs now in force among them, relative to the descent and conveyance, of property."

[31]Alvord (1909), 512.

[32]Burton, Stocking, and Miller (1922), 162ff. Settlers outside of Detroit were predominantly French and better insulated from British influence.

[33]Banner (2000).

much more lightly than Illinois, Indiana, and Michigan. For example, the first permanent settlement at Green Bay included just eight settlers in 1745. By 1766, with the area now under British control, there were a few French families at the fort and a few French families living nearby. Most of the land grants were located in Green Bay or in Prairie du Chien. The surviving evidence does not show that civil-law courts were operational in Wisconsin. The origin of the Ohio land grants listed in table 2.2 is a mystery.[34] Hence Ohio and Wisconsin are classified as common-law states.

Table 2.4 shows the beginning and ending dates for civil law for Illinois, Indiana, Michigan, and Missouri. In the last column, the duration of civil law, the first number is calculated using the date of acquisition as the end of civil law, and the second number is calculated using the date in the third column as the end of civil law. In comparison to the mid-Atlantic states, the Midwestern states had civil law for longer, and the civil-law legal regime ended at date that was much closer to statehood.

The Lower South

As a result of French, Spanish, English, and American territorial ambitions, the states in the Lower South—Alabama, Arkansas, Florida, Louisiana, and Mississippi—had contact with a number of governments.[35]

Classification of Louisiana and Arkansas as civil-law states is straightforward. As tables 2.2 and 2.3 show, Louisiana had both a large number of land grants and a large preacquisition population. It was governed by Spain and France continuously up to acquisition. Finally, the operation of the civil-law courts is well documented.[36] Like Louisiana, Arkansas was continuously governed by Spain and France. Arkansas was, however, a lightly populated frontier region, with relatively few land grants and low population. Morris Arnold's legal history of Arkansas indicates that courts in Arkansas operated in a manner similar

[34]We could not find evidence that they were French. Several works mention a congressional donation to French settlers who were swindled out of their money by the Scioto Company. Congress appears to have donated about 25,000 acres, which is substantially less than the amount listed in table 2.2.

[35]This discussion is based on Pickett and Owen (1900) and Brewer (1872) for Alabama; Herndon (1922) and Ashmore (1984) for Arkansas; Fairbanks (1901) and Tebeau (1972) for Florida; Barbe-Marbois (1830) and Cummins, Wall, and Schafer (2002) for Louisiana; Lowry and McCardle (1891) and Bettersworth (1959) for Mississippi.

[36]See Dargo (1975) and Fernandez (2001).

to the courts in Illinois and Indiana.[37] The military commander or a notary, if one was available, handled routine legal cases and forwarded complicated and high-valued cases to New Orleans for resolution.

Discussion of the legal history of Florida, Alabama, and Mississippi is complicated by the fact that the British controlled the region encompassed by all three states from 1763 to 1779, when the Spanish governor of Louisiana captured Natchez. In 1780, the Spanish captured Mobile. Some part of the territory further east remained in British hands until 1783, when Florida returned to Spanish control. In 1795, a boundary dispute was settled, and the parts of Alabama and Mississippi north of the thirty-first parallel were ceded to the United States in the Treaty of San Lorenzo. This left Spain with Florida and a narrow strip of Alabama and Mississippi along the Gulf Coast. This strip included Biloxi and Mobile.[38]

Of these three, classification of Florida as civil law is the simplest. Florida was under Spanish control with the exception of a twenty-year period of British control from 1763 to 1783. From then until 1821, the state was under Spanish control. Most of the settlement dates from this second period, since nearly the entire Spanish population evacuated upon British acquisition and the British did the same twenty years later. Thus, even if one dates civil law as beginning in 1783 in table 2.4, Florida still had civil law for the thirty-eight years immediately prior to American acquisition.

Mississippi and Alabama are also classified as civil law. It is worth reviewing their histories in some detail. Both states were lightly inhabited, and the population was clustered in the southern parts of these states. It is not clear what law the British used between 1763 and 1779–1780 in the inhabited parts of the two states. In an effort to attract British settlers and in turn stabilize control, the British government appears to have implemented a more traditional colonial policy with British government and common-law courts. This effort to attract British settlers seems to have been modestly successful, although some British left when there was a transfer to Spanish rule.

The settlements at Natchez, Mobile, and Biloxi and on the Tombigbee River came under Spanish control in 1779–1780. In 1798, three years after the signing of the Treaty of San Lorenzo, the United States finally took possession from Spain of all land between 31 degrees and 32 degrees, 28 minutes latitude.

[37] See Arnold (1985).

[38] See Matthews (1987) on Florida. On Natchez, Mississippi, see Holmes (1963) and on Mobile, Alabama, see Hamilton (1910). For West Florida, see also *Archives of Spanish Government of West Florida, 1782–1816*.

Table 2.4.
Duration of Civil Law in Civil-Law and Selected Other States

State	*Approx. first permanent settlement*	*Approx. date of control by a common-law country*	*Approx. end of civil law*	*Statehood*	*Duration of civil law*
Mid-Atlantic					
Dutch New Netherland (DE, NY, NJ, PA)	1624	1664	1665	1787–88	40–41
The Midwest					
Illinois	1700	1765	1790	1818	65–90
Indiana	1732	1765	1790	1816	33–56
Michigan	1668	1760	1792	1837	92–124
Missouri	1735	1804	1816	1821	69–81
The Lower South					
Alabama	1702	1798/1813	1798/1813	1819	18–111

Arkansas	1686	1804	1816	1836	118–130
Florida	1565	1821	1822	1845	38–256
Louisiana	1715	1803		1812	88–present
Mississippi	1699	1798/1813	1798/1813	1817	19–114
The Populated West					
Arizona	1700	1848/1853	1864	1912	148–164
California	1769	1848	1850	1850	79–81
New Mexico	1700	1848/1853	1876	1912	148–176
Texas	1718	1836	1840	1845	118–122

Notes: All dates are approximate. Dates are taken from the state histories listed in the appendix to chapter 2 and were cross-checked against the online version of Encyclopedia Britannica and other sources. In column 2, where there are two numbers, the first number(s) include the date(s) of territorial acquisition by the British or Americans. The second number is the date that common law was adopted. In column 4, the dates range from the most conservative to the least conservative. The lower date is the difference between the time of settlement and territorial acquisition. The assumption is that territorial acquisition sufficiently changed civil law that even if it nominally continued, it was no longer truly civil. In the cases of Mississippi, Alabama, and Florida, historical evidence suggests that many foreigners left with the change of government. So the conservative dating begins with Spanish reacquisition of the territory, rather than original settlement.

The remainder came under United States control in 1813. The populations of Mississippi and Alabama in 1800 north of the thirty-first parallel were 7,600 and 1,250. Of these, roughly 4,660 people were located across the Mississippi River from Louisiana in Natchez (Mississippi). Nearly all of the 8,850 individuals listed in the Mississippi Territory in 1800 would have been governed by civil law up to 1798. The population of the territory south of the 31-degree line is not well documented, but appears to have been small. These individuals would have had civil law up to 1813.

As a result of their complicated histories, one can date control of Alabama and Mississippi by a common-law country as occurring in 1798 or 1813. A very conservative dating of civil law would be from 1779–1780 until 1798. These dates are used as lower bounds in table 2.4. Because both states spent substantial time under French and Spanish control during the eighteen century, including the period immediately prior to American acquisition, they are classified as civil-law states.

The Unpopulated West

The eleven states in the unpopulated West—Colorado, Idaho, Kansas, Montana, Nebraska, Nevada, North Dakota, Oklahoma, South Dakota, Utah, and Wyoming—do not have evidence of significant European settlement before American acquisition.[39] A variety of individuals did travel through the area. French and to a lesser degree Spanish and Mexican traders were active and in some cases maintained trading posts in various locations. However, the numbers of individuals involved were small, and there is no evidence of land grants or courts.

One state, Colorado, had six confirmed land grants totaling more than one million acres and so warrants further examination. The six grants were made to residents of New Mexico for colonization or ranching. There is limited evidence that the grants were actually occupied on a scale necessary to qualify as

[39]This discussion is based on Bancroft and Victor (1890) and Snook (1904) for Colorado; Hailey (1910) and McConnell (1913) for Idaho; Holloway (1868) and Arnold (1914) for Kansas; Bancroft and Victor (1890) and Malone, Roeder, and Land (1991) for Montana; Johnson (1880) and Olson and Naugle (1997) for Nebraska; Bancroft and Victor (1890) and Davis (1913) for Nevada; Lounsberry (1919) and Wilkins and Wilkins (1977) for North Dakota; Gibson and Harlow (1984) and Thoburn and Holcomb (1908) for Oklahoma; Robinson (1905) and Schell (1968) for South Dakota; Bancroft and Bates (1889) and Stout (1971) for Utah; and Bancroft and Victor (1890) and Coutant (1899) for Wyoming.

permanent settlement before American acquisition in1848.[40] Indeed, historical sources identify the first permanent settlement in Colorado as being established in 1851.[41] The lack of settlement on the land grants was in part attributable to the fact that four of the six were granted in the 1840s. Another issue was the presence in some locations of hostile Native Americans.

All eleven states are classified as common law in table 2.1.

The Populated West

It is relatively straightforward to classify the six states in the populated West—Arizona, California, New Mexico, Oregon, Texas, and Washington.[42] All six had permanent settlement and operational courts at the time of acquisition. Arizona, California, New Mexico, and Texas were acquired from Mexico and had civil-law legal systems. Oregon and Washington were acquired from Great Britain and had common-law legal systems.

Spain established permanent settlements in California, New Mexico, and Texas in the late seventeenth century. Consistent with this history, table 2.3 documents that all three had substantial preacquisition populations. Table 2.2 shows that California and New Mexico also had large numbers of land grants by the time they were acquired from Mexico in the mid-nineteenth century. Despite having a much larger population, New Mexico had roughly the same number of confirmed land grants. The federal government was very slow to resolve land grants there and ultimately applied a very harsh standard. As a result, relatively few grants were confirmed in New Mexico. In Texas, the Spanish and Mexican government had issued grants for 26.3 million acres.[43] The Republic of Texas, later the state of Texas, patented these lands. Hence, more land was patented to the holders of foreign land grants in Texas than in the rest of the populated West combined. Notably, all three had Mexican government officials and operational civil-law courts.[44]

Arizona was more marginal in the sense that it was less attractive to settlers, settled later, and far from the seat of (regional) government in New Mexico.

[40]For discussion of these land grants, see Colorado State Archives (2001).

[41]Ubbelohde, Smith, and Benson (2006), 53.

[42]This discussion is based on Bancroft and Oak (1889) and Sheridan (1995) for Arizona; Bancroft (1888) and Starr (2005) for California; Bancroft and Oak (1889) and Prince (1912) for New Mexico; Carey (1922) and Bancroft and Oak (1886) for Oregon; Morphis (1875) and Fehrenbach (2000) for Texas; and Meany (1909) and Bancroft and Oak (1886) for Washington.

[43]General Land Office (2009).

[44]Cutter (1995) and Langum (1987).

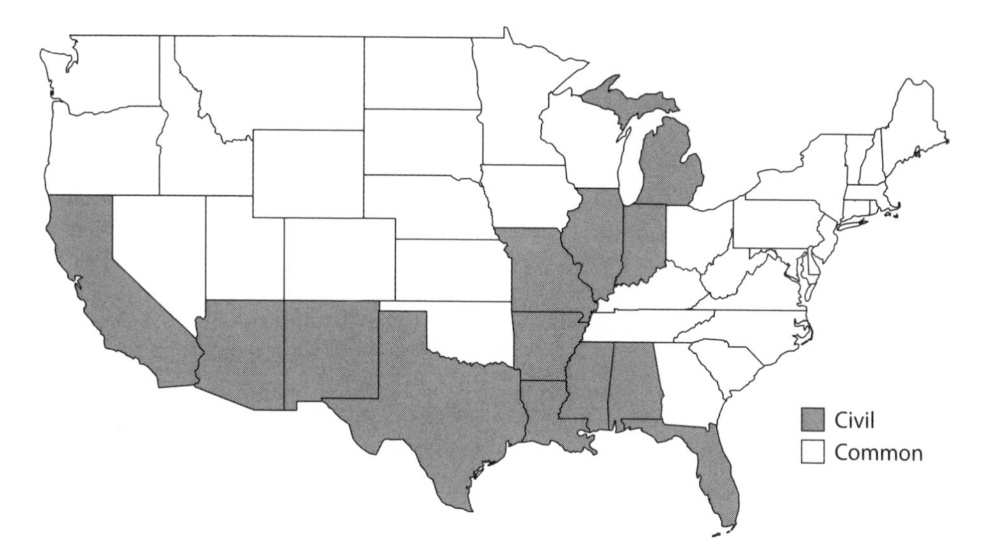

Figure 2.4. Civil-Law and Common-Law States.

This is reflected in the numbers of land grants and the preacquisition population estimates. While the number of settlers and land grants was small, Arizona was closely tied to New Mexico, both before and after American acquisition.

Oregon and Washington were under continuous British control from the time of settlement. Law enforcement was delegated to the Hudson Bay Company as part of its license. The 1821 Act of Parliament that permitted the issuance of licenses instructed the courts of Upper Canada "to take cognizance of all causes, civil or criminal, arising in any of the above mentioned territories, not within the limits of the provinces of upper or lower Canada, or of any civil government of the United States, and justices of the peace are to be commissioned in these territories, to execute and enforce the decisions of the courts, to take evidence, and commit offenders, and send them for trial to Canada."[45] For routine legal business, licensees could try criminal offenses and misdemeanors not punishable by death, and hear civil cases involving amounts up to two hundred pounds.[46] The Hudson Bay Company appears to have actually exercised this power.

[45] Quoted in Snowden (1909), 487–488.

[46] Gaston and Himes (1912), 166.

In sum, based on the historical evidence, table 2.1 classifies thirty-five states as common law and thirteen states as civil law. Figure 2.4 shows the location of these states.

Civil-Law Influence

Next we turn to how the balance of power between the legislature and the courts may have come to differ in common-law and civil-law states. One possible reason is that individuals with civil-law backgrounds were involved in constitutional conventions or the legislature or exerted indirect influence on these bodies. To the extent that such individuals preferred a more powerful legislature, they may have influenced aggregate views. A second reason is that individuals with common-law backgrounds may have preferred a stronger legislature or a more subordinate judiciary. Judges with both civil-law and common-law backgrounds may have been viewed with some suspicion. Judges with civil-law backgrounds may have seemed less competent to individuals with common-law backgrounds. At the same time, common-law judges, struggling with cases that dated to the civil-law period, may have seemed incompetent as well. Even if the judges seemed competent, individuals with common-law backgrounds, given the ability to choose, may have preferred a stronger legislature. In the following sections we document civil-law influence and discuss common-law reactions to civil law that may have influenced the relationships between legislatures and courts.

Population

If there are large numbers of individuals with civil-law backgrounds in a particular territory or state, then it is possible that they exerted either a direct or an indirect influence on the initial relationship between courts and legislatures. For reasons that are discussed shortly, their influence need not have been linear in population.

Early population estimates and the existence of land grants suggest civil-law influence, but they do not tell us about the composition of the population. That is, all of the individuals with French, Spanish, or Mexican heritage may have sold any land grants they held and left at the time of American acquisition. To examine the background of the population further, census data are used to address the following question: In 1850, in a given state, what fraction of men who were over the age of forty and held at least $100 in real property had civil-law backgrounds? The focus is on property-holding men over the age of forty because they were the individuals most likely to be influential in politics.

This exercise is less than ideal, because the states entered at different times. In particular, one would expect the percentages to be higher in Florida and the western states than in other civil-law states because they became part of the United States later. Unfortunately, the analysis is limited by the fact that the first census that asked about place of birth was taken in 1850.

Birth in civil-law areas includes individuals born in any civil-law state, as well as individuals who were born in France, Spain, Mexico, or French Canada. Because these men over forty were born in a civil-law area in 1810 or earlier, it is likely that they were exposed to civil law at least to some degree (see table 2.4 for the end of civil law in different states). The assumption is that many of these individuals arrived during the period in which civil law was in force or shortly thereafter. Thus, they would have played a critical role in the transmission of civil-law attitudes and norms.

Table 2.5 shows the percentage of the property-holding male population that had a civil-law background in 1850. The numbers for the Midwest were quite small, ranging from 2 percent in Michigan to 3.4 percent in Illinois. Given how early these states were acquired and how much in-migration occurred, the fact that there were any men with civil-law backgrounds is notable. Had it been possible to measure the same statistic in 1810 or 1820, dates that would be much more appropriate for measuring civil-law influence on the territory or the state, the share would surely have been larger. For example, contemporary reports indicate that Missouri was 40 percent French at the end of 1804.[47] In both Indiana and Illinois, difficulties with the Indians and with land titles delayed settlement until after 1800.[48] As a result, Illinois was largely French up to 1806. Indiana began to experience immigration slightly earlier.[49] Michigan was largely French into the 1820s.[50]

The percentages of men with civil-law backgrounds were extremely variable in the Lower South. Alabama had virtually no men with this background, while

[47]Bufus Easton, later territorial delegate to Congress from Missouri, in a letter dated at St. Louis, January 17, 1805, to President Jefferson, states that in 1801 the census taken of the inhabitants of Upper Louisiana showed a population of 10,301; and that according to the best-informed persons in the district the population at the close of 1804 had risen to over 12,000. Of this latter number he thought that two-fifths were French and the others mostly immigrants from the United States. (Copy of this letter in State Hist. Soc. of Missouri: original in MSS. Din., Library of Cong., Jefferson Papers, 2d Series, vol. 32.)

[48]Esarey (1915), 179.

[49]Alvord (1920), 414ff. discusses the reasons for this.

[50]See Dunbar and May (1995), 63 and the discussion in Lanman (1841) and Cooper (1920).

TABLE 2.5.
PROPERTY-HOLDING MALE POPULATION WITH CIVIL-LAW BACKGROUNDS IN 1850

State	*Year acquired*	*Over 40 share civil*	*Over 40 add Germanic civil*
The Midwest			
Illinois	1763	0.034	0.227
Indiana	1763	0.026	0.135
Michigan	1763	0.020	0.161
Missouri	1803	0.032	0.120
The Lower South			
Alabama	1798/1813	0.005	0.043
Arkansas	1803	0.048	0.071
Florida	1821	0.158	0.158
Louisiana	1803	0.565	0.647
Mississippi	1798/1813	0.028	0.043
The Populated West			
Arizona	1848/1853	0.500	0.500
California	1848	0.143	0.250
New Mexico	1848/1853	0.150	0.150
Texas	1846/1848	0.055	0.151

Notes: Computed from the public use sample of the 1850 Census of Population for white men over the age of 40 with at least $100 in real property. For dates, see the notes to table 2.3.

more than half of the men in Louisiana had civil-law backgrounds. Mississippi's and Arkansas's shares were consistent with the Midwestern states. Despite differences in the timing of acquisition, Alabama, Mississippi, Arkansas, and the Midwestern states began to experience significant in-migration between 1800 and 1810. Alabama had a very small French and Spanish population and bore the brunt of immigration from adjacent slave states. The French and Spanish populations in Mississippi and Arkansas were somewhat more protected. Interestingly, in 1810, a newly arrived minister stated that Natchez, the largest city in Mississippi at the time, was populated by "a mongrel race compounded of French, Spanish, and Negro, with a slight sprinkling of Anglo-Saxon ruffians and outlaws."[51] In 1803, American officials estimated that Arkansas was one-third French.[52] Florida, consistent with its more recent acquisition, had a substantial number of men with civil-law backgrounds. Most were located in Pensacola and St. Augustine.[53] As in the Midwestern states, the timing of the census was less than ideal for measuring the influence of civil law in the Lower South. One would have wanted to have measured these shares in 1810 or 1820.

In the populated West the shares, not surprisingly, were higher. The shares of men with civil-law backgrounds were extremely high for Arizona. Notwithstanding the gold rush, they were substantial in California. The share in New Mexico was surprisingly low. This may reflect a variety of factors including possible underenumeration, unwillingness to disclose property, and the wealth of a small number of foreign traders. The numbers for Texas are consistent with the substantial migration to the state beginning in the 1820s. Estimates in 1834 suggest that about 20 percent of the population was Mexican.[54]

One other statistic of possible interest is included in the third column in table 2.5. The definition of civil-law birthplace is expanded to include countries with Germanic civil-law legal systems. Under this definition, the share of men with civil-law backgrounds was over 10 percent in ten of the thirteen states. The timing of the arrival of these men is unclear. If it was early enough, it may have reinforced any existing civil-law norms.

[51] Quoted in Clark and Guice (1996), 201.
[52] Perkins, Peck, and Albach (1854), 786.
[53] Manley, Brown, and Rise (1997), 27.
[54] Manchaca (2001), 178, 201.

Participation of Individuals with Civil-Law Backgrounds in Early Politics

The preceding section on population discussed the possibility that individuals with civil-law backgrounds directly or indirectly influenced politics. If the influence was direct, it may be possible to observe it by examining the backgrounds of individuals who were active in early legislative (territorial) councils, territorial legislatures, state constitutional conventions, or early state legislatures.

One thing to keep in mind is the considerable length of time between territorial acquisition and statehood for many states, including the civil-law states. Although a few states achieved statehood very rapidly, the interval between acquisition and statehood was more typically twenty to thirty years. With an interval of more than sixty years, Arizona and New Mexico represented the other extreme.

The four Midwestern states spent between eighteen and forty-one years in unorganized or territorial phases. Three of the states—Illinois, Indiana, and parts of Michigan—were administered jointly up to 1805. Their governments were separated with the creation of the territory of Michigan in 1805 and the territory of Illinois in 1809. This history is important, because attitudes towards the judiciary may have formed early, sending the three states down similar paths. Missouri was acquired slightly later and was originally part of the district of Louisiana.[55]

Pierre Menard, a French Canadian fur trader, rose to prominence in early Midwestern politics. He represented the Illinois Country in the Indiana territorial legislature from 1803 to 1809 and was president of the Illinois territorial council from 1812 to 1818. On admission of Illinois to statehood, he served as lieutenant governor under Governor Shadrach Bond. Menard was related to the Chouteau family of Missouri.[56]

Although Menard was the most famous, other individuals active in early Midwestern politics had civil-law backgrounds.[57] For example, Charles Chabert de Joncaire, a member of a prominent French family in Detroit, served as one of the twenty-two members of the House of the First Territorial Legislature of the Northwest Territory. Gabriel Richard, French priest, was Michigan's delegate to Congress in 1823. Laurent Durocher, a French Canadian born in Missouri and educated in Montreal, was a member of the Michigan territorial

[55]Louisiana was part of the district of Orleans. Ironically, the district of Louisiana was under the temporary jurisdiction of the Indiana Territory for a little less than a year in 1804–1805.

[56]See Alvord (1920) and Reynolds (1887).

[57]See Burton, Stocking, and Miller (1922), Parkins (1918), and Lanman (1841) for Michigan and Houck (1908) and Violette (1918) for Missouri.

council and of the 1835 constitutional convention. Nathaniel Pope moved from Kentucky to Missouri in 1804, just as it was transitioning from common law to civil law. Fluent in French, he was then appointed secretary of the territory of Illinois in 1809. Auguste Chouteau was one of nine members on Missouri's first legislative council in 1813. John Scott, a Missouri lawyer, represented the interests of French families in Ste. Genevieve on the territorial council. With the backing of French families in St. Louis, he won a hotly contested race to be the territorial delegate to Congress.

The five states in the Lower South moved from acquisition to statehood at a pace roughly similar to that of the Midwestern states. The discussion for Louisiana will be brief. The population numbers and the political history demonstrate the power of the French population. They refused to adopt common law, except for criminal cases, and despite his misgivings regarding the situation, Thomas Jefferson was ultimately forced to accede to their wishes.[58]

Arkansas was attached to the district of Louisiana and then to the Louisiana Territory (later renamed the Missouri Territory) from 1804 to 1819. Thus, much of the discussion regarding Missouri applies here. The influence of the French in Missouri likely had some impact in Arkansas. Arkansas had virtually no representation in the Missouri territorial legislature because of its tiny population. In his history of Arkansas, Herndon (1922) writes: "When the [Arkansas] territory was organized in 1819 the English-speaking inhabitants were in the majority. Among the leading citizens at this time were: William O. Allen, Louis Barton, Charles Bogy, Jacob Bright, Stokeley H. Coulter, William Craig, John B. Dagler, Terence Farrelly, Mathias Foos, Peter (or Pierre) Le Fevre, Eli J. Lewis, Samuel Mosely, Frederick Notrebe, James and Hewes Scull, Etienne, Nathaniel and Stephen Vasseau, and the Vaugines."[59] Clearly there was some French presence. However, none of the names of the participants in the first territorial legislature are obviously French, nor were any of the participants explicitly identified as French or Spanish or Canadian.

Alabama and Mississippi were part of the Mississippi Territory from 1798 until Mississippi's admission as a state in 1817. There appears to be little evidence of individuals with civil-law backgrounds who were active in government in either state.

Florida had a greater civil-law presence.[60] Achille Murat, a well-known lawyer, was Napoleon's nephew. Arriving in 1821 or 1822, by 1824 he had

[58] See Dargo (1975) and Fernandez (2001).

[59] Herndon (1922), 410.

[60] See Fairbanks (1901) and Tebeau (1972).

purchased two large pieces of land, one near St. Augustine and a second near the territorial capital in Tallahassee. Eligius Fromentin, a French lawyer who had been practicing in Louisiana, was a territorial judge. Fromentin had a huge dispute with Andrew Jackson, the first territorial governor, and resigned after one year. The political power of individuals with civil-law backgrounds appears to have been strong. Florida sent Joseph Marion Hernandez to Congress in 1822–1823.

The most influential person in Florida with a civil-law background was Henry Marie Brackenridge. Brackenridge was briefly a lawyer and then was the judge for the Western District Superior Court from 1822 to1832. Born in Pittsburgh and raised in Ste. Genevieve in Missouri when it was under Spanish control, Brackenridge was fluent in French and Spanish. As a lawyer he spent considerable time in Maryland, Pennsylvania, Missouri, and Louisiana, before moving to Florida. In Louisiana, he had been deputy attorney general of the Territory of Orleans (Louisiana), and then a district judge. "Foreseeing . . . the difficulty under which the Territory would labor for the want of a suitable code of laws," Brackenridge "procured a volume of the digested legislative acts of Missouri, which had been similarly situated, having been also a province of Spain."[61]

The four civil-law states in the populated West had substantial numbers of Hispanic politicians. California and Texas bypassed the territorial stage entirely because of their large preacquisition populations and the substantial numbers of Americans who were already living there. In California, the number of Hispanic delegates to the state constitutional convention in 1849 was substantial, eight out of forty-eight. This may understate the civil-law contingent, because a number of other delegates were either born in Europe or had been longtime residents of California. Many of the latter had married into Hispanic families. Hispanic politicians were elected to the state legislature from 1849 to 1858.[62]

The 1836 Texas constitution was signed by three important Mexicans. Lorenzo de Zavala and Jose Antonio Navarro, both eminent Mexican politicians, were on the committee to draft the constitution. Perhaps not surprisingly, the constitution and early legislation bear a strong resemblance to Spanish and Mexican practice.[63] A third prominent Hispanic was José Francisco Ruiz, Navarro's uncle. De Zavala was the first interim lieutenant governor of the Republic of Texas. Navarro was the only Hispanic delegate to the Convention

[61] Manley, Brown, and Rise (1997), 23.

[62] *The State Register and Year Book of Facts* (1859), 194–200.

[63] See McKnight (1959).

of 1845. He helped write the first state constitution and successfully protected Hispanic rights. After statehood, he served two terms in the Texas Senate.

In New Mexico and Arizona, small populations led to very late admission. These states spent more than sixty years as territories.[64] New Mexico sent Jose Manuel Gallegos as its delegate to Congress in 1853, and it continued to send Hispanic delegates and members of Congress. Between 1848 and 1870 more than half of the delegates or territorial legislators were Mexican Americans. Moreover, Mexican Americans dominated the leadership positions.[65] Arizona, once it separated from New Mexico in 1863, had fewer Hispanic politicians, because of greater American in-migration. In 1864, just two of the nine members of the legislative council and none of the members of the territorial house appear to have been Hispanic.[66]

Land Grants

One possible link between foreign land grants and the balance of power between legislatures and courts is the uncertainty created by land grants. If territorial or early state common-law judges struggled with issues related to land grants, individuals with common-law backgrounds may have viewed legislative dominance as the solution. There is at least one example of effect. Governor Claiborne of the Mississippi Territory (and later of the Orleans Territory) wrote to James Madison from Natchez on December 20, 1801: "The Legislature is engaged on a new judiciary system. The manner in which the Superior and Inferior Courts have heretofore been arranged, is generally condemned. There is certainly room for improvement. One-half perhaps more, of the citizens, have no confidence in the Judges. The Legislature participates in this feeling and will, I fear, be inclined to legislate more against men than upon principles. It is an unpleasant state of things, and will be for me the source of much trouble. A late decision [on land grants] made by the Superior Court for this Territory has occasioned much complaint, and roused the sympathies of the Legislature. . . . I should be happy to have your opinion on the matter." In February 1802, he provided Madison with an update: "A violent dispute has sprung up between the Legislature and the Chief Justice, (who has many friends,) and they have taken steps to procure his removal or impeachment."[67]

[64] Vigil (1980).
[65] Gomez (2007), 89.
[66] Wagoner (1970).
[67] Claiborne (1880), 222–223.

In Missouri during the territorial period, many American officials and residents looked down on the existing legal system. In his legal history of Missouri, Stuart Banner put forward seven reasons for their reaction. He specifically mentions land grants as the basis for two of these reasons. Land grants led to a split in politics, with French and Americans who held land grants forming one group and other American residents forming the other. Further, legitimating the land claims required that one view the existing legal system as legitimate. Tied up in this conflict were self-interested views on who should hold wealth. While Banner does not specifically link land grants to the legislature's relationship with judges, the discussion suggests that such a link may have been present.

Adoption of Common Law and Hybridization

The timing of the adoption of common law and the extent of hybridization after the adoption of common law are also indirect indicators of civil-law influence. The Northwest Ordinance and later enabling legislation for territories east of the Mississippi typically specified that common law was in effect, although as previously mentioned, certain rights of French settlers were protected.[68] Thus, Alabama, Illinois, Indiana, Michigan, and Mississippi had common law from the beginning. In Michigan, however, there was some confusion regarding the default legal rules. In 1810 the territorial legislature of Michigan felt compelled to declare:

> Whereas, the good people of the Territory of Michigan may be ensnared by ignorance of the laws of other governments under which this territory has heretofore been, that is to say, of the *Coutume de Paris*, or common law of France, the laws, acts, ordinances, arrets and decrees of the ancient kings of France, and the laws, acts, ordinances, arrets and decrees of the governors or other authority of the province of Canada, and the province of Louisiana, under the ancient French crown and of the governors, parliaments or other authorities of the province of Canada generally, and of the province of Upper Canada particularly, under the British crown, which laws, acts, ordinances, arrets and decrees do not exist of record, nor in manuscript or print in this country and have never been formally repealed or annulled.[69]

[68] One of the most contentious of these rights was the right of French settlers to hold slaves, despite the ban on slavery in the Northwest Ordinance. This arose from the provision continuing "their law and customs now in force among them, relative to the descent and conveyance of property." Property was deemed to include slaves.

[69] Quoted in Kimball (1966), 309.

What the provision regarding common law meant was a significant point of controversy in some locations for quite a while, because common law itself varied between the United States and Britain and over time. Compounding the problem was a general lack of legal reference books on the frontier.

Congress took a slightly different approach in Missouri, Arkansas, Louisiana, and Florida. The language of the legislation for all four states allowed for the continuation of the laws that were in force prior to acquisition."[70] In 1807 in the Louisiana Territory (Missouri and Arkansas) American officials wrote that they wanted "to assimilate by insensible means, the habits and customs of the American and French inhabitants; by interweaving some of the regulations of the latter into our Laws, we procure a ready obedience, without violence or complaint."[71] Although American statutes and practices were adopted, common law was quite late in coming. It was only formally adopted by the Missouri territorial legislature in 1816. In Louisiana, which retained civil law, elements of common law crept in.[72] For example, the territorial legislature adopted common-law elements such as trial by jury and habeas corpus. Through land commissions, property was brought into the American system. Merchants in interstate trade adhered to the common-law norms of merchant communities elsewhere. Federal law, as it applied to Louisiana, was common law. In Florida, in contrast, one of the first actions that the territorial legislature took in 1822 was to adopt common law. So, with the exception of property and water issues, civil law disappeared.

In the New Mexico Territory, which originally encompassed New Mexico and Arizona, the language governing the choice of law was quite vague. Section 17 of the act to establish a territorial government for New Mexico states that "the constitution, and all laws of the United States which are not locally inapplicable, shall have the same force and effect within the said territory of New Mexico as elsewhere within the United States." The continued use of civil law in practice appears to have rested on the local inapplicability clause. In 1871, the New Mexico Supreme Court ruled that the civil law of Mexico was still in force.[73] In 1876, the territorial legislature formally adopted common law.[74] In the fall of 1864, shortly after the formation of the Arizona Territory, it adopted common law. In November 1864, Governor John Goodwin explicitly thanked the legislative council, noting: "Since its acquisition by the United

[70]Quoted in Friedman (1986), 168.

[71]Friedman (1986), 168, quoting Judge John Coburn to Secretary of State James Madison, 1807.

[72]See Friedman (1986), 171–176.

[73]Derden (1910), 81.

[74]Bakken (1974), 35–36.

States, the Territory has been almost without law or government. The laws and customs of Spain and Mexico had been clashing with the statute and common law of the United States, and questions of public and private interest had arisen, which demanded careful but decided action. These questions have been met and satisfactorily settled."[75]

California and Texas did not go through the territorial phase, so they needed to explicitly decide on the choice of law. In California, the Hispanic population was politically powerful enough to force a debate on the merits of retaining civil law at the state constitutional convention in 1849.[76] Elisha Crosby, chairman of the Judiciary Committee, observed, "There was quite an element of Civil Law in the Legislature and many wanted that adopted as a rule."[77] Although their efforts failed, it may have influenced views of the courts. Although Texas independence occurred in 1836, civil law continued to be in force until 1840. This continuation is suggestive of civil-law influence. Strikingly in 1845, Chief Justice John Hemphill, chairman of the Committee on the Judiciary for the Texas Constitutional Convention, wrote, "I cannot say that I am very much in favor of either chancery or of the common-law system. I should much have preferred the civil law to have continued here in force for years to come."[78]

In Texas, California, and New Mexico, and to a lesser extent elsewhere, pieces of civil law were explicitly incorporated into state law following the adoption of common law.[79] These included laws for marital property, wills and succession, and property rights in water. For example, community property, a civil-law tenet, survived in Texas, California, and New Mexico. It was then adopted by other western states.[80] Prior appropriation became the dominant doctrine with respect to water. All of these decisions represented choice of doctrine and so are suggestive of civil-law influence. One does not want to overstate the effects of civil law here, since these choices appear to have had an element of pragmatism. In California, continuation of community property seems to have hinged on the fact that women would be more willing to migrate if it were in force. And prior appropriation may have better suited the climate. Nevertheless, these choices represented a break with common law.

[75] Farish (1916), 127.
[76] Browne (1850), 258ff.
[77] Crosby (1945), 58.
[78] Quoted in Butte (1917), 699.
[79] See Worcester (1976).
[80] See van Kleffens (1968), Burns (2001), and McKnight (1996).

Complaints and Other Influences

In places where legal cultures came into conflict, complaints arose from parties on both sides. While the content of the complaints is fairly mundane, it is indicative of civil-law resistance and the extent to which individuals with common-law backgrounds looked down on the extant system. Although one response was to change the legal system, a related impulse may have been to try to assert legislative dominance and so restrict the power of judiciary.

The extent of the materials on the clash of legal cultures varies by state. Four states—Arkansas, California, Missouri, and Louisiana—have detailed legal histories that document the clash of legal cultures.[81] In Illinois and Indiana, perhaps because of French numerical dominance and the territorial structure, the opinions of the American officials and the French population were captured in petitions and official letters. Evidence of conflicting opinions has also survived in New Mexico and, to a much more limited degree, in Arizona. In some other locations, where clashes surely occurred, little evidence seems to have survived in the available written record. It may be that individuals in Florida were too afraid of Andrew Jackson to complain openly. French residents in Mississippi and Alabama may also have viewed open resistance to the new legal system as a poor strategy. In Michigan, four years of British common law may have limited complaints to the new American government. A brief check of Canadian sources did not turn up protests regarding the new legal system. In Texas, complaints may have been voiced to the Republic in 1840 when common law was adopted, but there is little record of them.

Individuals with civil-law backgrounds were often unhappy with the new (partially) common-law legal system. French residents in Illinois wrote to Congress that the local judge had denied "us, as we conceive, the right reserved to us by the constitution of the Territory, to wit, the laws and customs hitherto used in regard to descent and conveyance of property, in which the French and Canadian inhabitants conceive the language an essential."[82] The French population in Vincennes (Indiana) complained to the local judge that because of the imposition of common law, "laws are too complex, not to be understood and tedious in their operation."[83] A Missouri resident wrote to President Thomas Jefferson in 1805, "Many people here do not like the Change & every Law that

[81] See Arnold (1985), Banner (2000), Fernandez (2001), and Langum (1987).
[82] Alvord (1920), 405 quoting *American State Papers, Miscellaneous,* 1:151, 157.
[83] Friedman (1986), 169.

is pass'd puts them on a Worse Situation than they would have been under the Spaniards is Criticiz'd & the Worst Construction put on."[84]

Individuals with common-law backgrounds were scarcely more tolerant.[85] Amos Stoddard, who took possession of St. Louis for the United States, reported, "The laws, rules of justice, and the forms of proceedings were almost totally arbitrary—for each successive Lieut. Governor has totally changed or abrogated those established by his predecessor."[86] Lawrence Friedman notes that "American officials had no particular sympathy for the culture of French settlers. Judge John C. Symmes, who came to Vincennes in 1790, reacted to the French with chauvinistic disgust."[87] William Claiborne, who took possession of New Orleans for the United States, similarly described the legal system in New Orleans as "in most points incongenial with the principles of our own Government."[88] American merchants who lived in Mexican California continually complained about the weakness of the civil-law legal system and the low caliber of judges.[89] In New Mexico, similar concerns were voiced.[90]

A tendency towards legislative dominance may have been reinforced by a variety of other hard-to-document factors. The ideas embodied in the French Revolution, including those on the judiciary, had received currency among the political and economic elite. And at the same time, many members of the common-law legal elite were strongly influenced by (French) Roman civil law. Some of this was due to the Revolution and the subsequent publication of the Napoleonic Code in 1804. Quite apart from the current political philosophy or land grants, the elite may simply have viewed a stronger legislature as in their own self-interest. In common-law states, their ability to act on this view may have been much more limited that in civil-law states.

Obliteration and Populism

The prevailing wisdom among American legal historians is that common law obliterated civil law in the United States. Lawrence Friedman, a preeminent

[84]Banner (2000), 95 quoting J. B. C. Lucas to Thomas Jefferson, December 10, 1805, Lucas Papers, box 2, Missouri Historical Society. For additional examples, see Banner (2000), 94–95. See also Houck (1908), 380ff.

[85]For a discussion of the conflict in Canada when the Quebec Act was adopted, see Neatby (1937).

[86]Quoted in Bannner (2000), 89.

[87]Friedman (1986).

[88]Quoted in Banner (2000), 89.

[89]Langum (1987).

[90]Rosen (2001).

legal historian, writes, "A massive invasion of settlers doomed the civil law everywhere except in Louisiana."[91] His research and the work of other scholars on state legal systems supports this view. American lawyers quickly appeared and took positions on the bench or began to practice. They met little resistance, because the former colonies had very few lawyers. American lawyers brought with them the common-law approach to litigation. Even in Louisiana, common-law practices began to find their way into usage.

We agree that almost all of the laws and procedures associated with the previous civil-law legal systems were wiped out. Much ink has been spilled discussing the survival of small pieces of civil law in a small number of states. Our hypothesis is that having originally had a civil-law legal system influenced, through some of the many channels outlined above, the balance of power between the legislature and the judiciary.

In conversation, a number of legal scholars have argued that the striking empirical differences between civil-law and common-law states in their treatment of the judiciary are attributable to a different cause. They argue that civil-law states were (and perhaps are) more populist. In that view, populism is the driving factor.

To explore this hypothesis further, one first has to define populism. The most common association is with the Populist Party, which rose to prominence in the late nineteenth century. The party united western farmers in opposition to the gold standard. At its peak in 1892, the party presidential candidate, James Weaver, received 9 percent of the votes. Weaver carried four states—Colorado, Idaho, Kansas, Nevada—and received electoral votes from Oregon and North Dakota. None of these states were civil-law states. If one looks at state level officials, the story is similar—none of the ten governors that ran as Populists were from civil-law states. None of the six senators were from civil-law states, and only six of the forty members of the House of Representatives were from civil-law states. Three were from Alabama and three were from California. Clearly by this definition, the civil-law states were not populist.

The Progressive Party, formed in 1912 by a split in the Republican Party, had more overlap with the civil-law states. Thirteen congressmen were elected under the Progressive banner, representing California, Illinois, Louisiana, Michigan, New York, Pennsylvania, and Washington. Civil-law states are somewhat better represented than common-law states, but the evidence is hardly conclusive given that only four of the thirteen civil-law states elected congressmen from the Progressive Party.

[91] Friedman (1986), 168.

At least on its face, the evidence attaching the observed behavior to political movements appears weak. Even if the evidence had been strong, the question would have been the relationship among civil law, the relative power of the legislature and the judiciary, and the political movement. For example, populist or progressive political movements might have been reacting to a strong state legislature. The movement to elect state judges was a reaction to the power of the state legislature.[92] This change occurred in many civil-law and common-law states, and later states chose to elect judges when they entered the union. Although it had populist elements and was associated with Jacksonian democracy, partisan election of state judges is not the first thing that comes to mind when one says a particular state is populist. Further, conditional on having partisan elections, common-law states were much quicker to move away partisan elections than civil-law states, suggesting that there was something unique about civil-law states.

Conclusion

This chapter began by discussing why the balance of power between the state legislature and the state courts might affect economic and social outcomes, how the balance of power might have been influenced by colonial legal systems, and how the early balance of power might have shaped the evolution of state courts. Civil-law legal systems have stronger legislatures and less independent judiciaries. This may have led some residents in civil-law states to prefer a stronger legislature and a less independent judiciary and so to have influenced the early balance of power between the two branches of government. If the early balance of power shaped the subsequent evolution of state courts, then there should be systematic differences between civil-law and common-law states in the structure of their state courts.

Drawing on evidence of permanent settlement and the operation of courts, we classified states according to their colonial legal heritage. Thirteen states were classified as civil law, and thirty-five were classified as common law. An array of historical evidence was presented regarding possible civil-law influence during the interval between American acquisition and statehood. At the time of acquisition, most civil-law states had sizeable numbers of residents with civil-law backgrounds, substantial numbers of land grants, and at least some politicians with French or Hispanic roots. French and Hispanic resi-

[92] See chapter 5.

dents' complaints are suggestive of their cohesiveness and their perception of the likely outcomes. Congressional decisions not to impose common law in six of the eleven states that were territories and resulting delays in its adoption also speak to the power of the civil-law communities in those locations. Texas and California entered directly as states, but had sizeable civil-law communities. Had they taken the territorial route, the number of states with delays in adoption of common law would have been larger. No one piece of evidence is definitive, but taken together the available information suggests that individuals with civil-law backgrounds could plausibly have shaped early perceptions of the appropriate balance of power between the legislature and the judiciary. Residents with common-law backgrounds may have wanted a powerful legislature as well. The civil-law background of the state may have given them the political cover necessary to create a powerful legislature and a more subordinate judiciary.

CHAPTER 3

Initial Conditions and State Political Competition

THIS CHAPTER INVESTIGATES the relationship between five initial conditions in states—temperature, precipitation, distance to oceans, distance to rivers and lakes, and colonial legal system—and long-run levels of state political competition. State political competition is measured by examining the division of seats in the legislature between the political parties, although a number of other measures of state political competition are also examined. Figure 3.1 sketches some relationships between initial conditions and state political competition.[1]

We emphasize the importance of state political competition because it is thought to lead to better economic and social outcomes. Rodrik (1999) documented a strong positive association between the strength of democracy and manufacturing wages in countries around the world. He argues that the likely reason is that democracies have more robust political competition. In a detailed study of the Russian regions during the postsocialist transition, Remington (2010) finds a strong positive association between the extent of political competition and outcomes, including provision of public goods, tax compliance, and wages.

The relationship between political competition and economic and social outcomes in the United States has been the focus of considerable discussion, but little systematic analysis. The most obvious topic has been the low levels of political competition, slow growth, and poor social outcomes for some groups in the American South. Historians, economists, and political scientists have all written extensively about these issues. Some commentators, notably V. O. Key and Michael Holt, have discussed variation within the South in political competition and outcomes. Quantifying the impact of politics on economic and social outcomes is difficult, however, because of the potential for feedback

[1] With the exception of colonial legal system, the term "initial conditions" refers to conditions that were fixed (exogenous) features of a state prior to colonization. Some states were warmer or had more precipitation than other states, and this has changed relatively little over the last 500 years. Similarly, some states were located close to oceans, navigable rivers, or the Great Lakes, and other states were not.

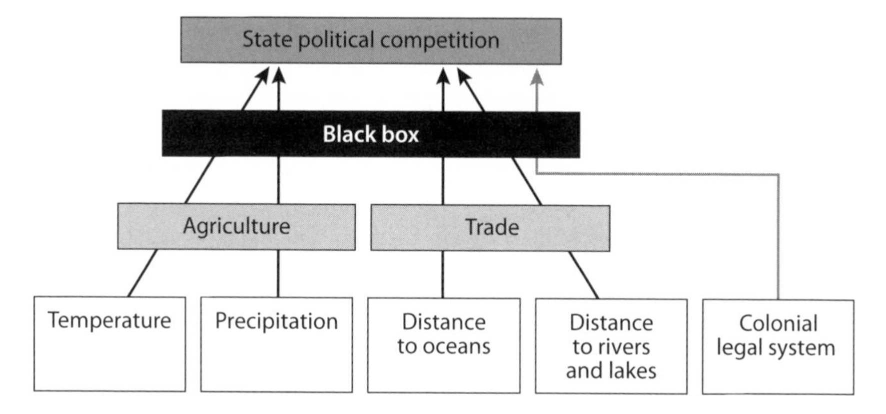

Figure 3.1. Initial Conditions and State Political Competition.

between politics and outcomes. Weak competition may cause poor outcomes, and poor outcomes may reinforce low levels of political competition.

Besley, Persson, and Sturm (2010) provide one of the first papers to show a causal relationship between state political competition and state economic outcomes. As a source of exogenous variation they use the 1965 Voting Rights Act, which forced many southern states to allow registration of practically all individuals of voting age. Higher levels of state political competition, as measured using vote shares for elected officials, led to three probusiness outcomes: lower state taxes, higher state infrastructure spending, and the increased likelihood of a state having a right-to-work law. These policies helped drive growth.

Political competition in American state legislatures has been highly persistent. This suggests that patterns were set early in a state's history. The historical literature frequently identifies agriculture and trade as competing political interests.[2] Tensions between these groups were at the heart of economic and political conflict in many states during many periods. For example, McGuire and Ohsfeldt (1989) found that merchants were more likely than farmers or other delegates to vote to ratify the federal Constitution at the thirteen state ratifying conventions held during 1787–1790, because it strengthened property rights, the enforcement of contracts, and policies related to trade. Main (1966) examined shifts in the number of merchants and farmers in six state legislatures around the time of the American Revolution. Ford (1984) explored

[2] For an overview of the historic determinants of suitability for agriculture and how they are quantified, see Motamed et al. (2009).

the dynamics of the transformation of the economy in the South Carolina up-country from 1865 to 1900 "to explain how tensions developed among planters, merchants, and yeoman [farmers] in the areas as a result of the economic metamorphosis."[3] Merchants and farmers were not always competing interests. For example, they came together to support the construction of plank roads in antebellum New York and to fight the railroad in postbellum Illinois. At the same time, their interests were distinct, and their cooperation was noteworthy.[4]

A colonial legal system could in theory also be related to state political competition, although the reasons why it would be are unclear.[5] Thus, it is included with a dotted line in figure 3.1. Empirically, it does not appear to be related to political competition.

Exactly how initial conditions or economic activities translated into political competition during the period 1866–2000 is not well understood, hence the black box in the figure. In the absence of a mechanism, it is difficult to discuss the impact of slavery and the Civil War on political competition, although their effects will be captured to some degree by initial conditions in a state. More systematic analysis of these issues is left to chapter 4.

State Political Competition

Because we present a broad overview of 150 years of American state political history, it is useful to address likely critiques by political historians. The first critique is that quantitative measures of political competition do not actually measure political competition.[6] For example, the division of seats is an imperfect measure of political competition for a number of reasons, most obviously because legislators do not always vote along party lines. Part of the issue is that political competition is inherently difficult to measure. By using a variety of common proxies for political competition and showing that patterns are broadly similar across different proxies, we hope to convince readers that the

[3] Ford (1984), 298.

[4] See Majewsky et al. (1993) and Woodman (1962).

[5] In the international context, countries with civil-law and common-law origins differ on a variety of dimensions. See Levine (2005) and La Porta et al.'s (2008) survey. In the U.S. context, however, with the exception of Louisiana, civil-law states adopted common law around the time of statehood.

[6] See Holbrook and Van Dunk (1993) and the discussion of the Ranney index.

proxies capture important aspects of state political competition. Two additional critiques are addressed in the next section.

The Ranney index is a commonly used measure of the extent to which one party dominates a state legislature. Several versions of the Ranney index exist, some of which include the party affiliation of the governor or multiply the shares of the seats held by different parties. For simplicity, we use an additive version of the index that excludes the governor:

$$\text{Ranney index} = 100 - (\text{abs}[(\text{percent Democrats in upper house}) + (\text{percent Democrats in lower house}) - 100]).$$

The political environment is most competitive when two parties each have 50 percent of the seats in both chambers. In this case, the Ranney index equals 100. Similarly, the political environment is least competitive when one party holds 100 percent of the seats in each chamber. In this case, the Ranney index equals 0.[7]

Although data on the party affiliation of state legislators are available as far back as the 1830s, these data are available for a larger number of states and years after the Civil War.[8] Table 3.1 presents summary statistics for the Ranney index. During the period 1866–2000, Illinois and Arkansas had the highest and lowest average values of the Ranney index at 82 and 11.

Figure 3.2 illustrates the striking difference in the evolution of the average Ranney index in the North and the South. Following the Civil War, the average Ranney index fell in both the North and the South. The fall was much greater in the South, where state legislatures came to be dominated by the Democratic Party. After 1960, the average Ranney index in the South began to grow rapidly. By the end of the twentieth century, the average Ranney index in the North and the South had converged.

Figure 3.3 presents some initial evidence of persistence in the state levels of political competition over time. It shows the average correlation in the

[7] One criticism of this version of the Ranney index is that it ignores whether legislative houses are controlled by different parties. An alternative Ranney index can be computed that accounts for whether or not the two state legislative houses are divided:

$$\text{RanneyALT} = 100 - \text{abs}[\text{percent Democrats in the upper house} - 50 + \text{percent Democrats in the lower house} - 50].$$

The correlation between our Ranney index and this alternative Ranney index is 0.97. So the two indexes tell similar stories.

[8] For a more detailed discussion of the data and its sources, see the appendix to this chapter.

TABLE 3.1.
SUMMARY STATISTICS FOR POLITICAL COMPETITION

	Ranney, 1866–2000	*Citizen voting, 1880–2000*	*Legislative professionalism, 1935–2003*
Average	51.7	80.5	0.16
Average North	60.4	85.6	0.17
Average South	23.5	63.4	0.13
Maximum	82.1	92.9	0.05
Minimum	11.2	43.4	0.42

Sources: See the appendix to this chapter.

Notes: The Ranney and Citizen voting indexes are for 36 states. The data for legislative professionalism are for 47 states.

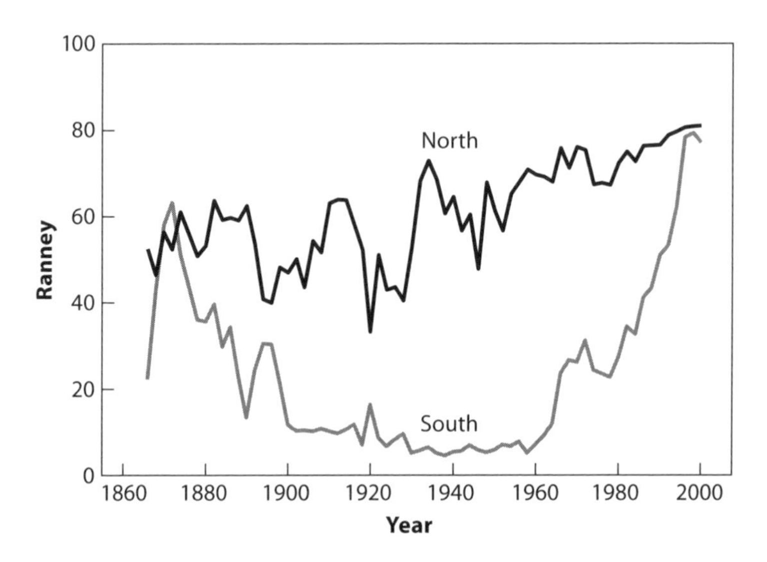

Figure 3.2. Evolution of the Ranney Index in the North and South, 1866–2000.

This figure uses the 36 states that have data for 1866–2000. Nebraska is dropped because it has a unicameral legislature for most of 1866–2000. Sources: See the appendix to this chapter.

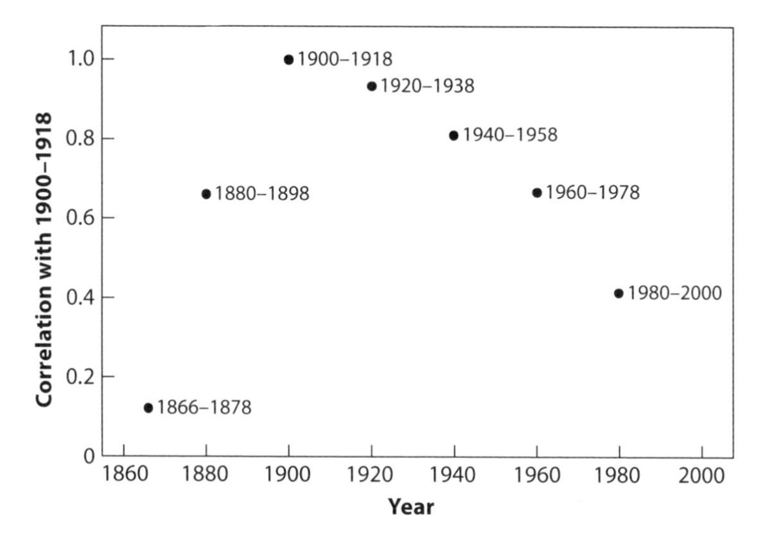

Figure 3.3. Correlation of State Ranney Indexes across Twenty-Year Intervals, 1866–2000.

This figure uses the 36 states that have data for 1880–2000. The Ranney index competition was averaged for each state over each specified time interval. The correlations were computed between the baseline period (1900–1918) and all other time intervals.

Ranney index across twenty-year periods, where the base period is 1900–1918. The "long panel" includes the thirty-six states that have political data covering 1866–2000. The "full" panel includes the forty-seven states that have political data for 1900–2000. The correlation for state Ranney indexes is low for the 1866–1878 period, probably because of the political turmoil associated with the Civil War and Reconstruction. The correlations for both the long panel and the full panel were high—0.48 or better—for 1880–1898, 1920–1938, 1940–1958, and 1960–1978. States with above average political competition in the baseline period also had above average competition in other periods.[9] The correlations finally fell to about 0.4 in 1980–2000.

The high correlation in state political competition over five twenty-year periods is quite remarkable. States' populations grew; individuals migrated into the state from other states and from foreign countries and migrated out of the state; the mix of economic activities changed; and yet states with high levels

[9]The correlation was substantial within the North and within the South as well. The correlation was somewhat higher and lasted for longer in the South.

of political competition in 1900–1918 had high levels of political competition in earlier and later years. Some degree of correlation is to be expected. What is surprising is the high levels of correlation over a long period.

An alternative measure of political competition is citizen voting. Voters tend to make their choices along party lines for low-profile "down-ballot officers" such as the attorney general and secretary of state. Vote shares substantially greater than 50 percent for one party indicate weak political competition. The data on citizen voting are available beginning in 1876. The citizen voting index is similar in construction to the Ranney index:

citizen voting = 100 – (2 * abs[(votes for Democratic candidates in broad elections) – 50]).

The possible values range from a low of 0 to a high of 100. During the period 1876–2000, Indiana and Mississippi had the highest and lowest values of competition at 93 and 43.

Figure 3.4 illustrates the evolution of citizen voting in the North and the South. The values of citizen voting imply that 62 percent to 86 percent of the citizens in an average southern state were voting Democratic from 1880

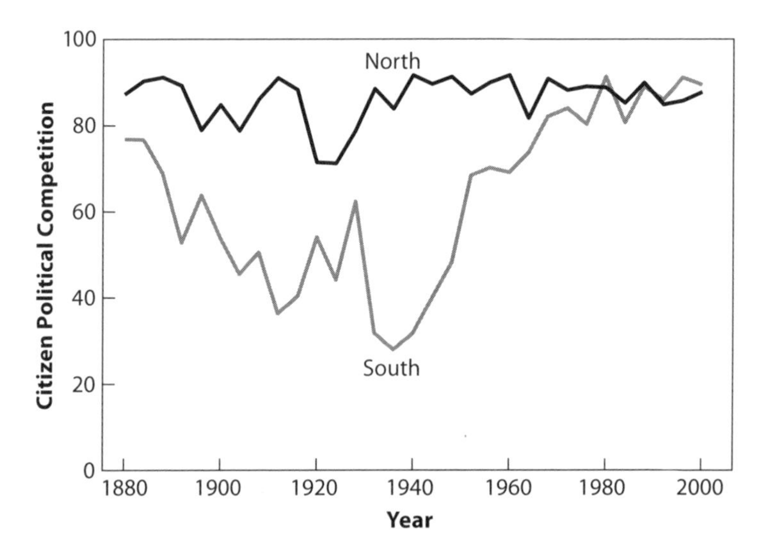

Figure 3.4. Evolution of Citizen Political Competition in the North and South, 1880–2000.

This figure uses the 38 states that have data for 1880–2000. Sources: See the appendix to this chapter.

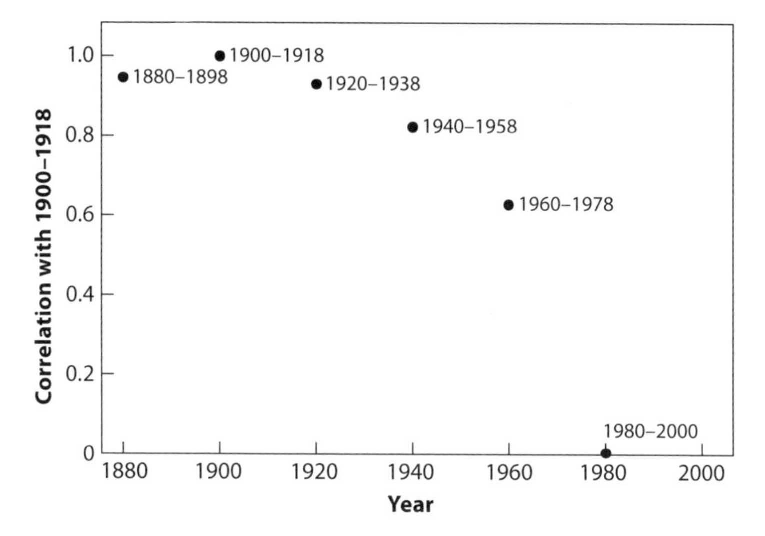

Figure 3.5. Correlation of State Citizen Competition across Twenty-Year Intervals, 1880–2000.

This figure uses the 38 states that have data for 1880–2000. Citizen political competition was averaged for each state over each specified time interval. The correlations were computed between the baseline period (1900–1918) and all other time intervals.

through the 1950s. Republican voters' low numbers and lack of concentration meant that the state legislatures had very few Republicans.

A comparison of figures 3.2 and 3.4 reveals other notable differences between the evolution of the Ranney index and citizen voting. First, citizen voting exhibited higher levels of political competition. Second, although both measures fell in the South after the Civil War, citizen voting in the South converged to northern levels around 1970. Convergence for the Ranney index took nearly three decades longer.

Like the Ranney index, citizen voting was highly persistent over time. Figure 3.5 shows the average correlation in citizen voting across twenty-year periods, where the base period is 1900–1918. From the 1880s through 1960–1978, the correlation never fell below 0.6. During the period 1980–2000, the correlation fell to approximately 0. Despite the many changes to state population, citizen political competition changed relatively little over the period 1880–1978.

Legislative professionalism may also affect political competition, in part because of its effects on who chooses to become a politician. The Squire index of legislative professionalism compares the averages for pay, staff size, and

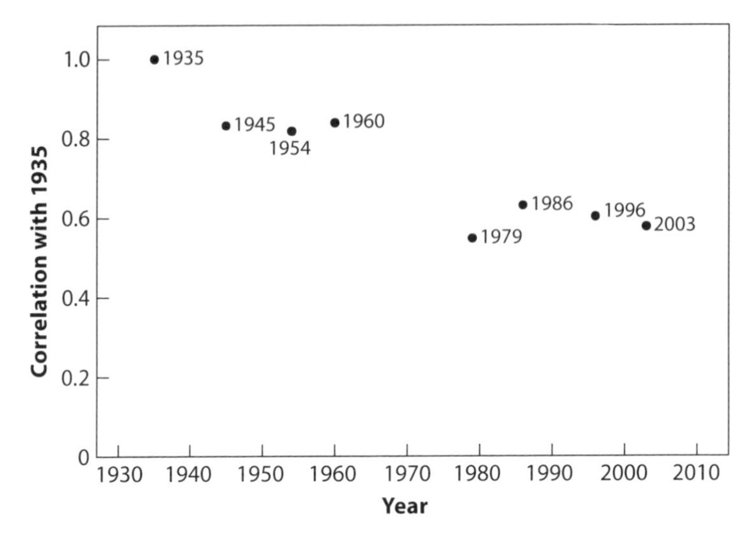

Figure 3.6. Correlation in State Legislative Professionalism, 1935–2003.

The correlations were computed between the baseline year (1935) and all other years. The full sample of 48 states is included.

number of days in session of state legislators with their counterparts in the U.S. Congress at nine points in time. "In essence, the measure shows how closely a legislature approximates these characteristics of Congress on a scale where 1.0 represents perfect resemblance and 0.0 represents no resemblance."[10] When the Squire index is close to 0, state legislatures meet infrequently and have small staffs. Massachusetts and Wyoming had the highest and lowest average values for legislative professionalism at 0.42 and 0.05.

Figure 3.6 presents evidence of persistence in legislative professionalism from 1935 to 2003. Over the period, the correlations with 1935 were always greater than 0.5 and frequently higher. The degree of correlation fell during the 1960s and 1970s as many states' governments—including legislatures—became more professional. Since 1979, however, the correlation appears to have stabilized at around 0.6.

Table 3.2 shows the correlations among the three measures of political competition. The Ranney index is strongly correlated (0.88) with citizen political competition. The Ranney index's correlation with legislative professionalism is positive, but more modest (0.30).

[10] Squire (2006), 4.

TABLE 3.2.
CORRELATIONS AMONG MEASURES OF POLITICAL COMPETITION

	Ranney, 1866–2000	*Citizen voting, 1880–2000*
Ranney, 1866–2000	1.00	
Citizen voting, 1876–2000	0.88	1.00
Legislative professionalism, 1935–2003	0.30	0.20

Note: Correlations are for the 36 states for which all 4 measures are available.

The analysis in this section highlights the extent to which all three measures of political competition were persistent over time, particularly during the period 1880–1978. This persistence is striking in light of the many changes that states underwent during this period.

INITIAL CONDITIONS

The literature on growth and institutions suggests a number of initial conditions that may be relevant for the evolution of political institutions in the United States. This section begins by discussing the four initial conditions relating to agriculture and trade. It then discusses other initial conditions that were not used in our analysis.

Over the nineteenth and twentieth centuries, the relative importance of agriculture has varied across states. One reason is variation in temperature and precipitation. In general, places with very short growing seasons, low precipitation, or both are not well suited to agriculture. For example, Illinois is inherently better suited for agriculture than Maine, which is colder, or Arizona, which is much drier. A second reason is that historically high-value export crops such as cotton, sugar, tobacco, and rice required high temperatures and substantial precipitation to grow.

Both Engerman and Sokoloff (1997) and Acemoglu, Johnson, and Robinson (2001) argue that climatic conditions affect political institutions. They offer different mechanisms—slavery and disease—but the main point is that both approaches link climate to politics. Historiography in the United States strongly links slavery to politics as well. Our interests are somewhat broader,

TABLE 3.3.
SUMMARY STATISTICS FOR INITIAL CONDITIONS

	Temperature (Fahrenheit)	*Precipitation (inches/mo.)*	*Distance to navigable rivers and Great Lakes (km)*	*Distance to oceans (km)*	*Civil*
Average	51.92	2.94	512.7	380.1	0.27
Standard deviation	7.84	1.13	365.8	325.0	0.45
Maximum	70.59	4.74	1,324.6	1,189.1	1.0
Minimum	39.44	0.73	101.8	16.7	0

Sources: see the appendix to this chapter.
Note: All 48 continental states are included.

in that we are interested in the competition between agricultural interests and trade interests in both the North and the South.

Thus, temperature and precipitation are used as initial conditions for agriculture. Detailed state-level data on precipitation and temperature are available for every state for the period 1895 to 2000. These long-term averages are used as the initial conditions for temperature and precipitation.[11] Descriptive statistics are presented in table 3.3. Average state temperature ranged from 39 to 71 degrees, and average monthly precipitation ranged from 0.7 to 4.7 inches. Table 3.4 shows that temperature and precipitation are moderately correlated (0.52).

Additional measures of suitability for agriculture such as soil quality do exist. One problem is that they were measured at later periods and tend to be more affected by human activity. Thus, they are less suited to being used as initial conditions.[12]

Given the primacy of the Civil War as an explanation for the subsequent evolution of political competition, one might argue that membership in the Confederacy should be used instead of temperature or precipitation. There are at

[11] See the appendix to this chapter for data sources and a more detailed description of the data.

[12] Measures of soil quality are also available. Soil quality, however, has changed over time in part because of the activity of farmers. See Lal (1999).

TABLE 3.4.
CORRELATIONS AMONG INITIAL CONDITIONS

	Temperature	*Precipitation*	*Distance to navigable rivers and Great Lakes*	*Distance to oceans*
Temperature	1.00			
Precipitation	0.52	1.00		
Distance to navigable rivers and Great Lakes	0.18	–0.57	1.00	
Distance to oceans	–0.19	–0.65	0.42	1.00
Civil	0.54	0.13	0.23	–0.03

Note: All 48 continental states are included.

least two reasons not to use the Confederacy as the primary measure. Membership in the Confederacy was endogenous. Political leaders in individual states made a variety of choices along the path leading up to the war that determined whether the state would be a member of the Confederacy or the Union. Had a different set of historical contingencies occurred, Missouri, Arkansas, Kentucky, Tennessee, West Virginia, Virginia, and Maryland might have ended up on a different side of the war; or a political compromise might have been reached that avoided the war altogether. In addition, using the Confederacy misses important variation within the North and the South in suitability for agriculture and in political trajectories. Nevertheless, it is worth noting that temperature and precipitation are correlated (0.75 and 0.55) with membership in the Confederacy.

Access to water transportation is commonly used as a proxy for the extent to which internal and external trade emerges. Water offered a cheaper means for moving goods than roads and was the dominant mode of moving goods prior to the spread of railroads. Adam Smith wrote:

> As by means of water carriage a more extensive market is opened to every sort of industry than what land carriage alone can afford it, so it is upon the sea-coast, and

> along the banks of navigable rivers that industry of every kind begins to subdivide and improve itself, and it is frequently not till a long time after that those improvements extend themselves to the inland part of the country.[13]

Access to water was likely to affect the composition of economic activity. In the international context, Easterly and Levine (2003) show that being landlocked is negatively associated with the quality of political institutions.

The first measure of access, distance to an ocean, captures access to external trade. Oceans were commonly used to move goods to and from foreign countries. Atlantic trade with Europe was dominant, although merchants in California both before and after American acquisition traded extensively with merchants in Hawaii and South America. Oceans were also used to move goods for internal purposes, such as from Charleston to Boston or from New Orleans to New York. But even there, some of the movement was of goods to major ports that would then be shipped internationally.

The second measure of access, distance to a major navigable river or the Great Lakes, captures access to internal trade. These water sources facilitated the movement of goods among the regions of the United States. At the most aggregate level, agricultural goods were moved east, and processed goods and imports were moved west. Access to internal trade clearly also supported external trade, in the sense that it allowed movement of goods to major ports for export.

Distances were computed using data from Rappaport and Sachs (2003). They have county-level measures of distances in kilometers to large or medium-size ocean ports, to large or medium-size ports on the Great Lakes, and to a navigable river.[14] Our measure of a state's distance to ocean is the average distance to the nearest large or medium-size ocean port for all counties in the state. Similarly, our state distance to a navigable river or the Great Lakes was constructed by computing the average of the shortest distance to a navigable river and distance to the nearest large or medium-size port on the Great Lakes for each county in the state. This measure was averaged over all counties in the state to get the average state distance to rivers and lakes.

Although canals were important during the nineteenth century, they are not part of our distance measures because they were constructed.[15] Canals were

[13] Smith (1914), 16.

[14] For more details on construction of the distances to internal and external water, see the appendix to this chapter.

[15] One can debate whether navigable rivers were constructed as well, since man-made improvements often increased their navigability. But these navigable rivers were almost always navigable without the improvements.

built in places where value added was high, the terrain was suitable, and individuals could induce state legislatures or private parties to support construction. One could create relatively exogenous measures by examining canals that were planned but not built or by examining geographic measures of potential suitability for canals.[16] Because most planned canals were adjacent to navigable rivers, the Great Lakes, or the ocean, their actual impact on distance to these water sources is likely to have been small.[17]

Table 3.3 shows that average distances to a navigable river and the Great Lakes and to the oceans varied widely. Table 3.4 provides the correlations among the initial conditions. Precipitation, average distance to navigable rivers and Great Lakes, and distance to oceans are all moderately correlated. This reflects that fact that locations far away from water tend not to have large amounts of rain.

Two other potential initial conditions are widely discussed in the American or international context. The first, political culture, was constructed by Daniel Elazar.[18] Although Elazar (1966, 1984) calls his variable "political culture," his classification is based on the ethnicity and religion of settlers. Thus, it can be interpreted as measuring culture more broadly. Elazar offers a detailed eight-category classification, but these types can be thought of as variants on three political cultures—moralistic, individualistic, and traditionalistic. He describes them as follows:

> Since individualistic political culture emphasizes the centrality of private concerns, it places a premium on limiting community intervention—whether governmental or nongovernmental—into private activities to the minimum necessary to keep the marketplace in proper working order. . . . In the moralistic political culture, individualism is tempered by a general commitment to utilizing communal—preferably nongovernmental, but governmental if necessary—power to intervene into the sphere of "private" activities when it is considered necessary to do so for the public good or the well-being of the community. . . . Traditionalistic political culture is rooted in an ambivalent attitude toward the marketplace coupled with a paternalistic and elitist conception of the commonwealth. It reflects an older, precommercial attitude that

[16] Fogel (1964) includes maps of actual and proposed canals. See figure A-1 (between pages 250 and 251).

[17] The correlations between distance to canals and distance to Great Lakes, distance to navigable rivers, and distance to oceans in our data set are 0.95, 0.49 and 0.37, respectively.

[18] Other competing classifications exist. However, as Lieske (1993) noted, only Elazar's classification has been widely used empirically. Although Elazar's classification was constructed to explain mid-twentieth-century political behavior, Berman (1988) showed that Elazar's classification has explanatory power for the Progressive Era.

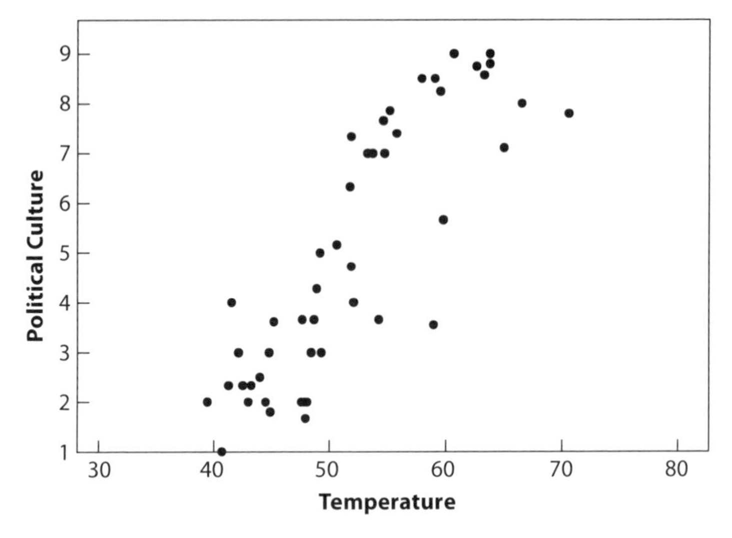

Figure 3.7. Average Annual State Temperature and Elazar's State Political Culture.
Political culture is scaled from 1 to 9, with 1 being the most moralistic and 9 being the most traditionalistic. For sources see the appendix to this chapter. The correlation between political culture and temperature is 0.847.

> accepts a substantially hierarchical society as part of the ordered nature of things, authorizing and expecting those at the top of the social structure to take a special and dominant role in government.[19]

Despite the possible importance of political culture, two factors preclude its use. Like membership in the Confederacy, political culture may be endogenous. In a different historical contingency in which different groups initially populated different areas and different migration streams followed, one might have had different political cultures arise. But even if culture were exogenous, Sharkansky's (1969) translation of Elazar's measure of political culture onto a numerical scale is highly correlated (0.85) with temperature. Figure 3.7 shows the relationship between the temperature and political culture, which is approximately linear. The linearity makes it difficult to disentangle the effects of temperature and political culture. Given that only one measure can be used, we use temperature because it is exogenous while political culture is not. The ef-

[19] Elazar (1984), 94–99.

fect of temperature on political competition can be thought of as capturing both suitability for large-scale agriculture and the associated political culture.

Within the literature on initial conditions, a substantial subliterature exists that examines the effect of natural resources on a variety of outcomes.[20] Here we ignore natural resources for reasons of endogeneity and timing. Although the deposition of minerals and oil thousands of years ago was exogenous, their discovery and development depended on factors such as increases in population and the development of demand for the natural resources. With respect to timing, oil and mineral discoveries would not happen until the second half of the nineteenth century and often later. At the time of major discoveries, patterns of state political competition had been established.[21]

State Political Competition and Initial Conditions

We begin by addressing two additional critiques by political historians. The first is that everyone already knows that political competition has been persistent and was shaped by initial conditions. Certainly, one piece of the story is well known to political historians. A large number of scholars have examined the North-South split in state politics that occurred as a result of the Civil War.[22] The degree of persistence across different measures of state legislatures and citizen voting may be new to some political historians. Further, the effect of other initial conditions on state legislatures and state political competition more broadly has received virtually no attention. Therefore, some of our results may be new to political historians.

The second critique is that state political competition at any time is determined by culture, religion, class, race, or other aspects of the composition of

[20] A line of research on natural resources and growth began with Sachs and Warner (1995, 1997). A related line of research on natural resources and political outcomes began with Ross (2001). Haber and Menaldo (2009) find that the results linking natural resources to nondemocratic regimes do not hold up in time series. Mitchener and McLean (2003) find a positive relationship between share of the workforce in mining in 1880 and income per worker in the United States.

[21] Goldberg et al. (2008) find that discoveries of natural resources tended to preserve existing American state political structures.

[22] There are many studies, and we will mention just a few. The seminal example of southern state political history is V. O. Key's *Southern Politics in State and Nation* (1949). An important example of the study of sectional state politics is Michael Holt's *The Political Crisis of the 1850s* (1983). At the national level, an important example is Keith Poole and Howard Rosenthal's *Congress: A Political-Economic History of Roll Call Voting* (1997).

the populace.[23] These influences are surely important, but state political competition is quite persistent over time. Thus, it seems likely that initial conditions played a role.

The remainder of this section examines the extent to which initial conditions are related to state-level political competition. Both for simplicity and because a mechanism has not yet been specified, the focus is on the relationships between initial conditions and long-run levels of political competition. Given the high degree of persistence in levels of political competition observed in the previous section, the relationships are likely to be related to short-term levels as well.

Average temperature and precipitation were negatively related to state-level political competition over the period 1900–2000. Figure 3.8 plots the relationship between state temperature and the average value of each state's Ranney index. Figure 3.9 shows the analogous picture for precipitation. The Ranney index was lower in states with higher temperature and rainfall.

Both of our measures of distance to water transportation—distance to navigable rivers and Great Lakes and distance to oceans—were negatively related to state-level political competition, although in the case of oceans the relationship was extremely weak. Figures 3.10 and 3.11 illustrate these associations. The Ranney index was lower in states that were far from internal and external water transportation.

Legal origin could be related to political competition, although in the United States it is not clear exactly why this relationship would hold. Figure 3.12 shows the relationship between legal origin and the Ranney index. The average level of the Ranney index in the civil-law states was lower than in the common-law states, although the difference is not statistically significant. Common-law states were relatively tightly grouped, while the interquartile range for the civil-law states was very wide. This reflects the fact that two civil law states, Illinois and Arkansas, had the highest and lowest values of the Ranney index.

To investigate the effect of initial conditions on the Ranney index, the following regression was estimated:

$$\text{Ranney}_i = \alpha_0 + \alpha_1 \text{logtemp}_i + \alpha_2 \text{logprecip}_i + \alpha_3 \text{logdist rivers lakes}_i + \alpha_4 \text{logdist oceans}_i + \alpha_5 \text{civil}_i + u_i \quad (1).$$

[23] See Baum (1984), Benson (1961), Bourke and Debats (1995), Formisano (1983), Kleppner (1970), Kruman (1983), Levine (1977), and Maizlish (1983). See the discussion in McCormick (1986) and Formisano (1999) on the ethno-cultural view of politics and the earlier literature on the elite and class-based politics. See also Patterson and Caldeira (1984) and King (1989), who relate the Ranney index to state-level characteristics such as education, income, and population.

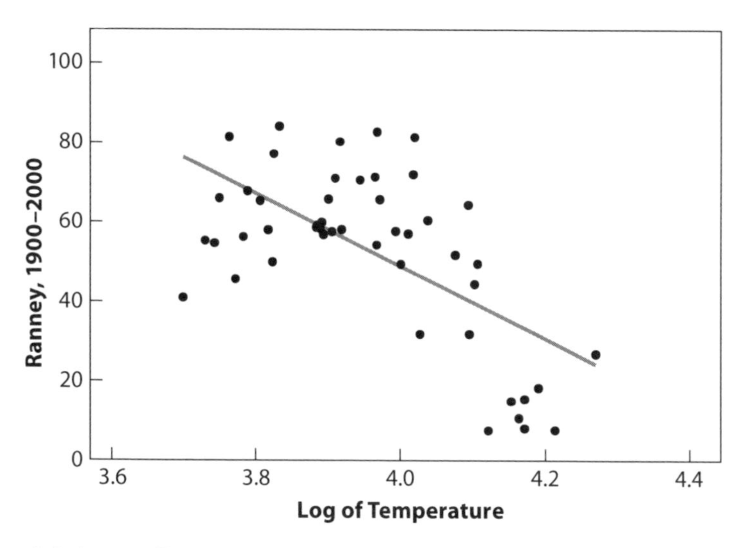

Figure 3.8. Average Temperature and the Ranney Index, 1900–2000.

Temperature is annual average temperature. Observations are for the 48 continental states. The *R*-squared for the regression of the Ranney index, 1900–2000 on the log of temperature that includes a constant is 0.393. Sources: See the appendix to this chapter.

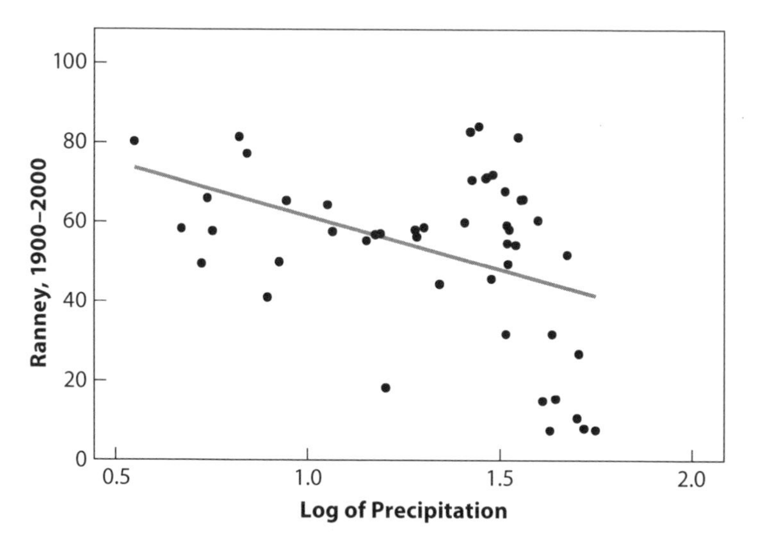

Figure 3.9. Precipitation and the Ranney Index, 1900–2000.

Precipitation is monthly average precipitation. Observations are for the 48 continental states. The *R*-squared for the regression of the Ranney index, 1900–2000 on the log of precipitation that includes a constant is 0.151. Sources: See the appendix to this chapter.

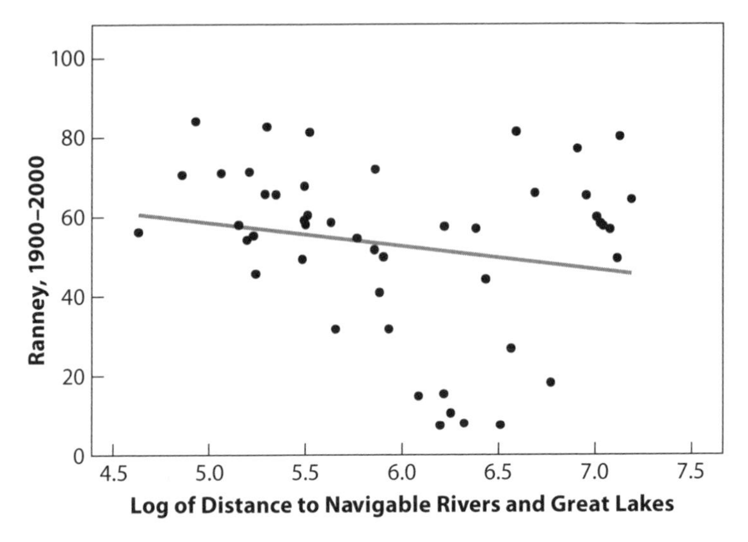

Figure 3.10. Distance to Navigable Rivers and Great Lakes and the Ranney Index, 1900–2000.

Average distance is measured in kilometers. The *R*-squared for the regression of the Ranney index, 1900–2000 on the log of distance to navigable rivers and great lakes that includes a constant is 0.039. Sources: See the appendix to this chapter.

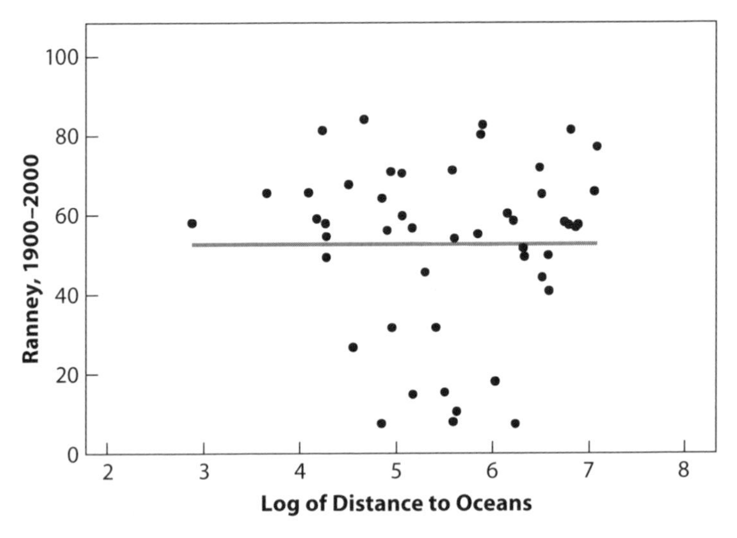

Figure 3.11. Distance to Oceans and the Ranney Index, 1900–2000.

Average distance is measured in kilometers. The *R*-squared for the regression of the Ranney index, 1900–2000 on the log of distance to oceans that includes a constant is 0.000. Sources: See the appendix to this chapter.

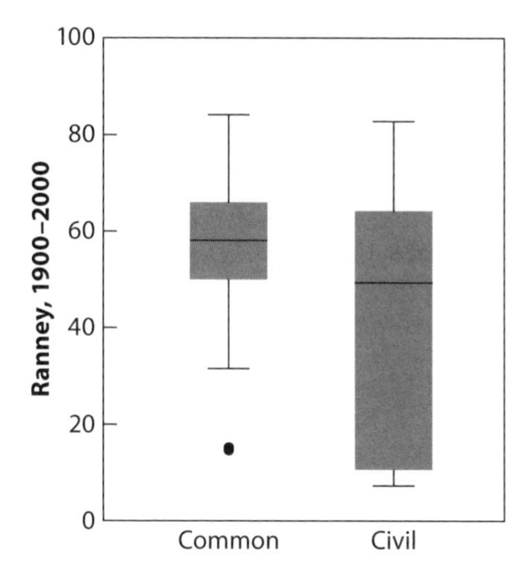

Figure 3.12. Legal Origin and Ranney Index, 1900–2000.

In each box, the central bar is the median state, the box contains the interquartile range and the T-sticks contain the entire population when no state is more than two standard deviations above or below average. When a state is more than two standard deviations above or below the average, it is depicted with a black circle that lies outside the T-sticks. Sources: See the appendix to this chapter.

In this cross-section, logtemp, logprecip, logdist rivers lakes, and logdist oceans are the logs of temperature, precipitation, distance to navigable rivers and the Great Lakes, and distance to an ocean. *Civil* is an indicator variable, which equals 1 if the state was classified as civil law in chapter 2, and u_i is a stochastic error term.

The regressions in table 3.5 indicate that, of the five initial conditions, two—log of precipitation and log of distance to internal water—were statistically significantly related to political competition. Further, the effects were sizeable. Over the period 1866–2000, a 10 percent increase in precipitation was associated with a 4.8-point fall in the Ranney index, while a 10 percent increase in the distance to internal water was associated with a 1.7-point fall in the Ranney index. States with higher levels of precipitation and states farther away from navigable rivers and the Great Lakes had lower levels of political competition.

Tables 3.6 and 3.7 show that this finding extends beyond the Ranney index. Citizen political competition and legislative professionalism were also lower in states with high levels of precipitation and states farther away from navigable

TABLE 3.5.
INITIAL CONDITIONS AND THE RANNEY INDEX

Dependent variable	*Ranney, 1866–2000*	*Ranney, 1900–2000*
Log of precipitation	–48.14***	–36.06***
	(16.05)	(12.55)
Log temperature	–11.13	–39.21
	(34.20)	(26.17)
Log of distance to internal water	–16.60**	–9.352*
	(6.449)	(4.870)
Log of distance to ocean	–3.421	–3.772
	(3.362)	(3.104)
Civil	0.853	–2.572
	(7.811)	(7.615)
Constant	277.5**	333.2***
	(102.2)	(83.71)
Observations	36	47
R^2	0.507	0.474

Notes: The notations *, **, and *** denote significance at the 10 percent, 5 percent, and 1 percent levels. Nebraska is dropped because it has a unicameral legislature for most of the years during 1866–2000. Standard errors are heteroskedasticity robust.

rivers and the Great Lakes. States more distant from oceans were also less professional.

A number of points can be taken away from the analysis thus far. The first is that some initial conditions—precipitation and distance to rivers and lakes—appear to be related to long-run levels of state political competition. Further, these relationships are not unique to the Ranney index. They hold across a variety of measures of political competition. The results are also consistent with the initial conditions acting through agriculture and trade, although they may have been acting through other channels as well. Finally, civil law appears not

Table 3.6.
Initial Conditions and Citizen Political Competition

Dependent variable	*Citizen competition, 1880–2000*	*Citizen competition, 1916–2000*
Log of precipitation	–21.41*	–17.53**
	(11.01)	(7.152)
Log temperature	–15.60	–14.54
	(23.37)	(15.75)
Log of distance to internal water	–8.678**	–5.671**
	(3.690)	(2.635)
Log of distance to ocean	–0.669	–1.450
	(1.799)	(1.375)
Civil	–2.684	–1.713
	(4.939)	(4.039)
Constant	227.0***	204.2***
	(70.77)	(50.23)
Observations	38	48
R^2	0.461	0.414

Notes: The notations *, **, and *** denote significance at the 10 percent, 5 percent, and 1 percent levels. Standard errors are heteroskedasticity robust.

to have been related to long-run levels of state political competition. While the result is not surprising, it was useful to demonstrate it.

State Constitutions

This section examines state constitutions, which provided the basic framework of state government, to provide additional evidence on the relationships between initial conditions and aspects of state government and on persistence. Unfortunately, many dimensions of state constitutions are difficult to quantify

TABLE 3.7.
INITIAL CONDITIONS AND LEGISLATIVE PROFESSIONALISM

Dependent variable	*Legislative professionalism, 1935–2003*
Log of precipitation	–0.046**
	(0.017)
Log of temperature	0.018
	(0.015)
Log of distance to internal water	–0.041***
	(0.015)
Log of distance to ocean	–0.046**
	(0.018)
Civil	0.038
	(0.027)
Constant	0.148***
	(0.012)
Observations	48
R^2	0.338

Notes: The notations ** and *** denote significance at the 5 percent and 1 percent levels. Standard errors are heteroskedasticity robust.

in a meaningful way. We examine four quantifiable dimensions: the length of the first state constitution, the length of the state constitution in 1990, the amount of particularistic content, and the number of seats in the state legislature.[24] The first two variables are measured in words.

Particularistic content refers to the composition of state constitutions. State constitutions include two types of provisions—framework provisions and statutory laws. Framework legislation covers governmental principles, processes, and institutions. Unlike framework legislation, statutory laws are not observed

[24] See Friedman (1988), Lutz (1994), and Tarr (1992). For an important discussion of how changes in state constitutions related to changes in the political and economic environments, see Wallis (2005, 2006, and 2008).

in the federal Constitution and are simply laws that have been upgraded to constitutional status. Hammons (1999) calls statutory laws particularistic legislation. He offers some examples of particularistic provisions: "All telephone and telegraph lines, operated for hire, shall each respectively, receive and transmit each other's messages without delay or discrimination, and make physical connections with each others lines, under such rules and regulations as shall be prescribed" (Oklahoma, Article 9, Section 5, 1907). "The people hereby enact limitations on marine net fishing in Florida waters to protect saltwater finfish, shellfish, and other marine animals from unnecessary killing, overfishing, and waste" (Florida, Article 10, Section 16, 1968).

The number of seats in the state legislature was typically set at the time the first constitution was written and then persisted over time. Figure 3.13 highlights this persistence.

Table 3.8 presents summary statistics on state constitutions. Along every dimension there is enormous variation across states. The longest initial constitution was nearly sixty times the length of the shortest one. By 1990, the longest current constitution was still nearly thirty times the length of the shortest one. Particularistic content ranged from 4 percent of the constitution to 73 percent. The average number of seats in the state legislature ranged from 48 to 428.

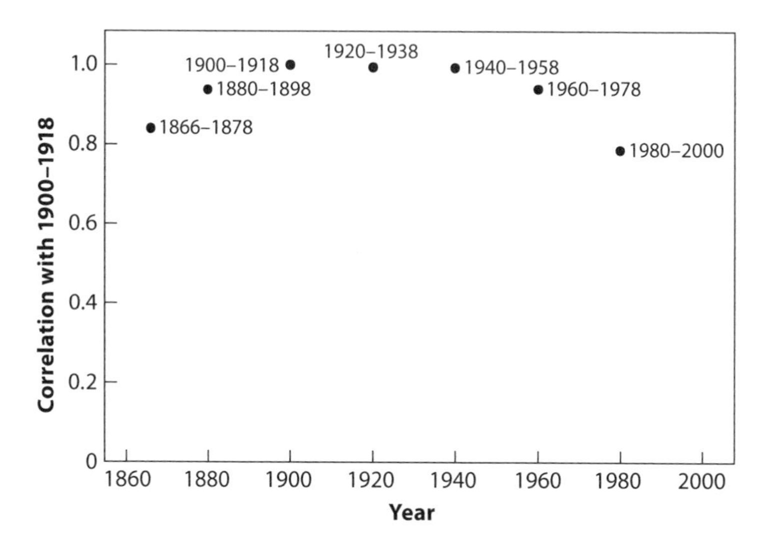

Figure 3.13. Correlation of Number of Seats in State Legislatures across Twenty-Year Intervals, 1866–2000.

This figure uses the 36 states that have data for 1866–2000. The number of seats was averaged for each state over specified time intervals. The correlations were computed between the baseline period (1900–1918) and all other intervals.

TABLE 3.8.
SUMMARY STATISTICS FOR STATE CONSTITUTIONS

	Initial length of constitution (in words)	*Length of constitution in 1992*	*Particularistic content in 1997–98*	*Seats in state legislature, 1900–2000*
Average	11,356	28,780	0.31	150.6
Average North	12,668	24,645	0.27	139.7
Average South	6,941	42,688	0.41	153.9
Maximum	58,200	174,000	0.73	427.7
Minimum	1,065	6,600	0.04	47.9

Sources: See the appendix to this chapter.
Note: All 48 continental states are included.

Table 3.9 shows the relationship between initial conditions and state constitutions. The first column shows that the length of the initial constitution was positively and statistically significantly related to temperature. States with higher average temperatures had longer state constitutions. The most striking thing about the next two columns is the persistent effect of the length of the initial constitution on the length of the constitution in 1990 and on particularistic content. States with longer initial constitutions had longer constitutions in 1990 and more particularistic content in their constitutions in 1997–1998. The fourth column indicates that the number of seats in the state legislature was positively and statistically significantly related to precipitation. States with more precipitation had more seats in the state legislature.

CONCLUSION

This chapter documented that state political competition was persistent over long periods of time and across a variety of different measures of political

TABLE 3.9.
STATE CONSTITUTIONS AND INITIAL CONDITIONS

Dependent Variable	*Log of length of original constitution*	*Log of length of constitution in 1992*	*Particularistic content in 1997–98*	*Seats in state legislature, 1900–2000*
Log of precipitation	–0.148 (0.301)	0.207 (0.271)	–0.0196 (0.0730)	100.5* (51.43)
Log of temperature	1.788*** (0.628)	–0.728 (0.616)	0.0621 (0.132)	–183.7 (125.2)
Log of distance to internal water	0.0195 (0.106)	0.129 (0.118)	0.00676 (0.0300)	–10.85 (17.63)
Log of distance to ocean	0.0948 (0.0804)	–0.105* (0.0613)	0.00905 (0.0151)	3.325 (9.234)
Civil	0.167 (0.195)	–0.0102 (0.111)	0.0211 (0.0293)	–3.596 (19.70)
Log of length orig. const.		1.138*** (0.123)	0.134*** (0.0282)	
Constant	–4.737** (2.289)	2.327 (1.730)	–0.435 (0.338)	792.0** (385.9)
Observations	48	48	48	47
R^2	0.372	0.759	0.565	0.296

Notes: The notations *, **, and *** denote significance at the 10 percent, 5 percent, and 1 percent levels. The state of Nebraska is excluded in the last column because, unlike the other continental states, it had a unicameral legislature during 1935–1974. Standard errors are heteroskedasticity robust.

competition. State constitutions also exhibit high degrees of persistence. This suggests that key features of state political institutions were established early in a state's history.

The historiography for the United States and for other countries suggests that initial conditions related to agriculture and trade were likely to have played formative roles. We find evidence that initial conditions related to agriculture and trade—specifically precipitation and distance to rivers and lakes—were strongly associated with long-run levels of state political competition. Notably, a state's colonial legal system seems not to have been associated with its political competition.

Appendix: Data Description and Sources

Temperature and Precipitation: Data on mean annual temperature and mean monthly precipitation are from the National Climatic Data Center at the National Oceanic and Atmospheric Administration.[25] The NCDC describes the data as follows: "The state-wide values are available for the 48 contiguous States and are computed from the divisional values weighted by area. The Monthly averages within a climatic division have been calculated by giving equal weight to stations reporting both temperature and precipitation within a division."[26] The observations for temperature were corrected for time of observation bias as described in Karl et al. (1986). The annual state values for 1895–2000 were averaged to generate the initial condition.

Access to Water Transportation: Access to water transportation was constructed from Rappaport and Sachs's data (2003).[27] They constructed distances in kilometers to a wide variety of water sources from county centroids in 2000 using geographic information systems. We use three of their distance measures. The first measure is distance to the closest ocean, which is their WTR_DO234_{si} variable, where the subscript "si" denotes state "s" and county "i". The second variable is the distance to the closest Great Lake, which is their WTR_DL234_{si} variable. Both are distance to a small, medium, or large natural harbor.[28] The third variable is the distance to the river as identified by Fogel (1964) for 1890, which is their WTR_DFR_{si} variable. Each distance measure is averaged across counties to obtain state-level averages, denoted WTR_DO234_s, WTR_DL234_s, and WTR_DFR_s. A state's distance to the ocean is the variable WTR_DO234_s and a state's distance to internal water is the average of WTR_DL234_s and WTR_DFR_s.[29]

Civil Law: The construction of the civil-law variable was described in chapter 2.

[25] The data were downloaded from http://www1.ncdc.noaa.gov/pub/data/cirs/.

[26] The state data are described in http://www1.ncdc.noaa.gov/pub/data/cirs/state.README.

[27] The data were downloaded from Jordan Rappaport's web page http://www.kc.frb.org/home/subwebnav.cfm?level=3&theID=10968&SubWeb=10782.

[28] Rappaport and Sachs (2003) describe the construction of this measure as follows: "Ocean and Great Lakes coasts and county boundaries are based on the 1:1.25 million ArcUSA Map constructed and distributed by ESRI Corporation (www.esri.com). For each [modern] county, the ESRI software package ArcView was used to calculate the distance to the nearest shoreline from the county's centroid (a mathematical approximation of 'the center' of an irregular polygon" (39–40).

[29] The distances are almost identical if they are weighted by county land area.

Political Culture: Data on culture is from Sharkansky's (1969) codification of Elazar's (1966, 1984) classification of state political culture.

Ranney Index and Seats in the State Legislature: The Ranney index is constructed from Burnham's (1986) data for the years 1834–1985 and from data collected by Tim Besley from the *Book of the States* for the years 1950–2000.[30] For states in every election year, it lists the number of seats in the upper and lower house and the number of seats held by each party within each house. Most states elect parts of the state legislature either every year or every other year. A few states elect their state legislature every fourth year. The same data set is used to compute seats in the upper and lower house of each state legislature. Total seats in any year equal seats in the upper and lower houses.

There are a number of different Ranney indexes that measure slightly different things. Our focus is the state legislature, so the party affiliation of the governor is not included and an additional amount is not added if the same party controls both houses and the governorship. An additive rather than a multiplicative measure is used.[31]

Citizen Political Competition: Citizen political competition is based on the data that was processed and analyzed by Ansolabehere and Snyder (2002) and then subsequently updated. Most states consistently report this data every fourth year (presidential elections). Some states elect every two years, and a few have elections every year. However, the cleanest view of the data is every four years.

Legislative Professionalism: Data on legislative professionalism are taken from Squire (2006 and 2007). The Squire index compares the average pay, average staff size, and average number of days in sessions of a member of a state legislative body with his or her counterpart in the U.S. Congress. When the Squire index is close to 0.0, state legislators have relatively low and small staffs and they meet relatively infrequently. This can be associated with a culture where legislators are pressed to find alternative income sources and where they are poorly informed about technical aspects of issues.

State Constitutions: The data on the number of constitutions per 100 years, and the amendment rate are from Lutz (1994). Particularistic content is from Hammons (1999). The length of the state constitution in 1992 is from the *Book of the States*. To count the number of words in each initial state constitution, constitutions posted on the NBER/Maryland State Constitutions Project were downloaded from http://www.stateconstitutions.umd.edu/index.aspx.

[30] The Burnham data (ICPSR #16) are available from ICPSR, http://www.icpsr.umich.edu/cocoon/ICPSR/STUDY/00016.xml. Tim Besley provided us with these more recent data.

[31] The additive and multiplicative measures are highly correlated.

Table 3.1A.
State Initial Conditions

State	*Precipitation (inches)*	*Temp. (F)*	*Civil*	*Distance to internal water (km)*	*Distance to ocean (km)*
Alabama	4.47	63.32	1	518.4	278.1
Arizona	1.06	59.77	1	1233.2	559.6
Arkansas	4.10	60.70	1	490.3	511.8
California	1.86	58.95	1	1324.6	126.6
Colorado	1.33	44.86	0	1000.9	1189.1
Connecticut	3.76	48.43	0	197.4	58.9
Delaware	3.70	54.70	0	249.2	68.0
Florida	4.50	70.59	1	708.5	93.7
Georgia	4.17	63.82	0	499.8	245.4
Idaho	1.57	43.98	0	1047.9	671.8
Illinois	3.15	51.90	1	199.1	363.2
Indiana	3.33	51.77	1	182.3	263.7
Iowa	2.68	47.60	0	279.3	499.0
Kansas	2.29	54.20	0	591.6	953.2
Kentucky	3.94	55.73	0	247.0	469.1
Louisiana	4.74	66.57	1	669.6	126.4
Maine	3.56	41.23	0	319.2	70.9
Maryland	3.58	53.68	0	239.9	71.0
Massachusetts	3.56	47.66	0	242.4	64.2
Michigan	2.59	44.48	1	172.3	70.5

Table 3.1A. (*continued*)

State	*Precipitation (inches)*	*Temp. (F)*	*Civil*	*Distance to internal water (km)*	*Distance to ocean (km)*
Minnesota	2.17	40.70	0	186.1	344.4
Mississippi	4.57	63.83	1	555.0	266.8
Missouri	3.40	54.59	1	351.4	656.3
Montana	1.28	42.12	0	727.7	907.6
Nebraska	1.90	48.70	0	502.1	886.7
Nevada	0.73	49.23	0	1247.1	356.2
New Hampshire	3.54	43.22	0	242.4	89.3
New Jersey	3.73	52.07	0	208.9	37.8
New Mexico	1.12	53.25	1	1144.0	975.2
New York	3.24	45.19	0	137.6	105.3
North Carolina	4.13	59.07	0	376.3	223.1
North Dakota	1.45	39.44	0	358.8	722.0
Ohio	3.16	50.66	0	127.9	145.2
Oklahoma	2.83	59.53	0	621.7	673.4
Oregon	2.24	48.09	0	1186.0	173.9
Pennsylvania	3.32	48.93	0	157.7	138.5
Rhode Island	3.59	49.34	0	244.0	16.7
South Carolina	4.00	62.64	0	438.8	175.6
South Dakota	1.53	44.76	0	366.2	716.3
Tennessee	4.34	57.93	0	349.6	552.6

TABLE 3.1A. (*continued*)

State	*Precipitation (inches)*	*Temp. (F)*	*Civil*	*Distance to internal water (km)*	*Distance to ocean (km)*
Texas	2.33	65.03	1	872.8	413.4
Utah	0.96	47.89	0	1127.9	847.3
Vermont	3.38	42.48	0	187.9	198.7
Virginia	3.55	55.15	0	286.3	140.6
Washington	3.09	47.98	0	1105.9	156.6
West Virginia	3.67	51.87	0	180.1	271.2
Wisconsin	2.61	42.96	0	101.8	134.1
Wyoming	1.10	41.53	0	803.3	1155.1

CHAPTER 4

The Mechanism

INITIAL CONDITIONS ASSOCIATED WITH trade and agriculture appear to related to state political competition. Why and how would these initial conditions have had a persistent influence on political competition? This chapter argues that these initial conditions shaped the occupational distributions of early state elites who, in turn, influenced the subsequent evolution of state political competition. Figure 4.1 provides a summary of the relationships among initial conditions, occupational homogeneity of the elite, and state political competition. Initial conditions shaped the early comparative advantage of the state economy. For example, temperature and precipitation influenced whether early elites came from agriculture or other occupations. Similarly, distance to water transportation influenced whether or not state elites engaged in trade. The mix of elite occupations influenced state political competition. When state elites worked largely in the same profession, a single party that reflected the interests of this occupation tended to dominate. When state elites worked in a broader mix of professions, different groups supported different parties, and political competition was stronger from the outset.

Alabama illustrates the issues that could arise from having relatively homogeneous elite. During Reconstruction, merchants were becoming a threat to the planter class, as merchants profited through the crop-lien system, and some invested their profits in land. Wiener describes this conflict in detail: "For six years after the conclusion of the war, Alabama's merchants and planters had competed as legal equals in the contest over control of the black tenants. . . . By 1871, with the effective end of Reconstruction in Alabama, the planter class apparently had regained enough political and economic power to mount a campaign to weaken the legal position of merchants holding crop liens."[1] Planters succeeded in passing a new antimerchant law "making their lien for supplies legally superior to that of the merchants."[2] Merchants' influence fell in the Black Belt, but conflict between the two groups continued. "In 1884 agrarian

[1] Wiener (1975), 86, 87.
[2] Wiener (1975), 88.

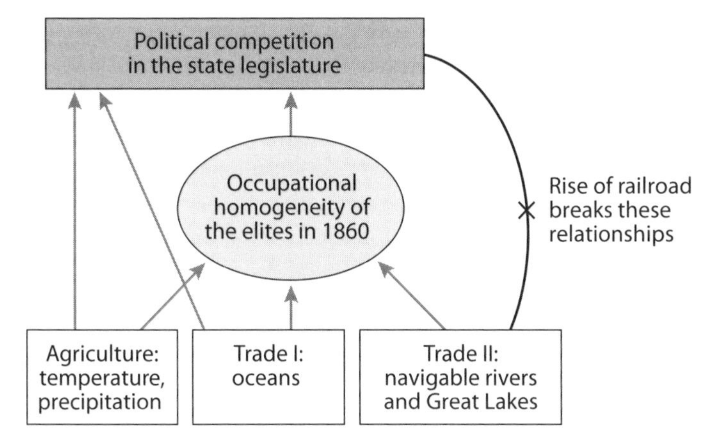

Figure 4.1. Initial Conditions, Occupational Homogeneity of the Elite, and State Political Competition.

radicals mobilized to demand repeal of all crop liens."[3] The planter-dominated legislature responded by eliminating merchant crop liens in most counties—while leaving landlord liens intact.

In a state with higher levels of political competition, it would be hard to pass or at least maintain legislation that overtly disadvantaged a particular group within the elite. Tamahana writes, "Late-nineteenth and early-twentieth-century contests among economic interests . . . shaped American law. . . . A loser in the legislature turned to the courts to re-fight the battle, and vice versa."[4] Courts during this period were becoming more activist. They began to use substantive due process and other doctrines to invalidate legislative decisions. Nelson finds that "Substantive due process was a weapon for invalidating any legislative act in which private interests, rich or poor, put the power of the government to use for their own ends at the expense of others."[5]

Occupational homogeneity exhibited a strong negative, persistent, and causal relationship with political competition in state legislatures over the period 1866–2000. We used data from the 1860 Census of Population to construct measures of the share of state wealth held by the state elite and occupational homogeneity of the state elite. Precipitation and distance to internal water transportation were related to occupational homogeneity, and temperature and

[3] Wiener (1975), 72.
[4] Tamahana (2006), 46.
[5] Nelson (1982), 154.

distance to oceans were related to wealth. Occupational homogeneity of the elite in 1860 exhibited a negative and strikingly persistent relationship with state political competition over time. In contrast, the share of state wealth held by the elite in 1860 exhibited a variable relationship to political competition over time. The question is whether the relationship between occupational homogeneity and state political competition is causal. The rise of the railroad dramatically changed the importance of distance to internal water transportation, and distance to internal water was strongly related to occupational homogeneity of the elite. This allows us to use distance to internal water transportation as an instrument to show that occupational homogeneity had a causal effect on political competition.

The levels of state political competition were highly persistent and strongly related to the occupational homogeneity of the early elite for a number of reasons. The geographical units within states that elected state legislators, usually counties, were for the most part dominated by one type of elite. County elites influenced who became a state legislator. Change in the underlying composition of economic activity was slow, and it took time for new elites to supplant old elites. Old elites may well have invested in maintaining their political power, even as their economic power waned. All of this meant that in the absence of substantial changes in the mix of economic activities or in party platforms, counties tended have legislators from the same party year after year. These patterns were reinforced by the fact that up to the 1960s there had been relatively limited reapportionment of seats to reflect changes in population. This last point is particularly important, because later representation tended to reflect the original distribution of representation and elites and not the current attributes of the population or the elite.

Finally, we explore the implications of slavery and the Civil War for our results. A concern is that occupational homogeneity is simply acting as a proxy for slavery or the South. This seems unlikely given that temperature and precipitation are allowed to influence state political competition both directly and through the occupational homogeneity of the state elite. The causal relationship between occupational homogeneity of the state elite in 1860 and subsequent state political competition remains strong, even controlling for slavery.

The Elite

Although the mechanism is one in which elites in different counties have different occupations and party affiliations, data constraints force us to use state elites as a proxy for local elites. Figure 4.2 illustrates how one would construct

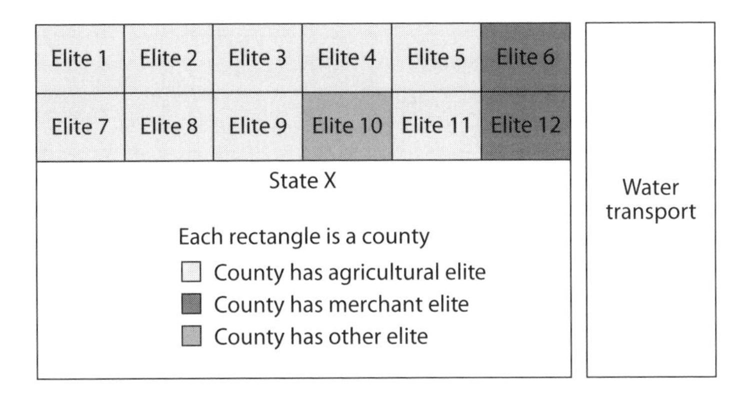

Figure 4.2. Occupational Homogeneity of the Elite Using County-Level Data.

occupational homogeneity if one had data that was detailed enough to examine elites by county. Unfortunately, the available data on elites, which will be described shortly, are not detailed enough to allow examination of local elites. Figure 4.3 illustrates how occupational homogeneity is constructed using state data. As long as the occupations of state elites are correlated with the occupations of local elites, they will be a good proxy for the occupations of local elites.

The state elite are defined as the top 1 percent of the wealth distribution of adult white males in the state, where adulthood is defined as beginning at age twenty-one. The state economic elite will likely overlap with and influence other types of elites, notably political elites. And measures of the top 1 percent of the economic elite will tend to correlate with measures of the top 0.5 percent and the top 2 percent. Thus, the measure of the wealth held by, and the occupational homogeneity of, the top 1 percent of the wealth distribution can be thought of as proxies for the wealth held by and the occupational homogeneity of the actual elite.

Of the early censuses, the 1860 census offers the best data for examining the elite. The 1798 census of housing values covers the small number of extant states. The next available wealth data are contained in the Censuses of Population for 1850, 1860, and 1870. The 1850 census asked only about real property. The 1860 census inquired about both real and personal property. Thus, relative to the 1850 census, it offered a more complete picture of wealth and covered a somewhat larger number of states. The 1870 census also inquired about real and personal property. The chaos of the Civil War, however, surely affected the wealth and the composition of elites in the North and the South. From 1870, there are no national data at all until the early twentieth century, when wealthy individuals began to pay income and estate taxes.

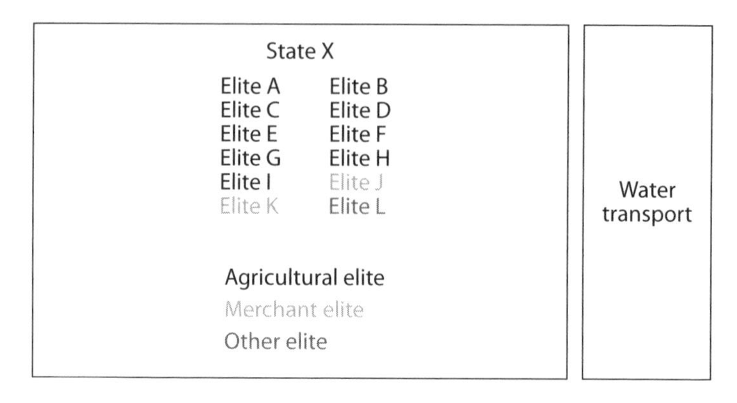

Figure 4.3. Occupational Homogeneity of the Elite Using State-Level Data.

The 1860 data do have their drawbacks, and it is useful to mention some of them here. These wealth data are unverified, self-reported data, with all the attendant problems of such data. The measures of real and personal property include only assets and not liabilities. As a result, they are imperfect proxies for wealth. A small amount of wealth is excluded, because women, children, and nonwhites are not included in the wealth distribution. At present, the publicly available data for 1860 is a 1 percent sample of the population. To ensure adequate sample size, states were required to have at least 600 adult white men in the 1 percent sample and thus at least 6 adult white men in the elite, which translates into having at least 60,000 adult white men in the state.[6] If a 100 percent sample were available, local elites could be examined, and a few more states could be included. As it is, the data cover only twenty-eight states. Finally, the measures of wealth and occupational homogeneity may have been somewhat influenced by the larger than usual cotton harvest in some southern states in 1859.[7] However, the effect on the occupational composition of the elite may not have been overly large, since in most states the occupational homogeneity of the top 1 percent and the next 1 percent was relatively similar. But wealth shares in the South may have been somewhat inflated, relative to what they would have been if the cotton harvest had been more typical.

[6]This requirement excluded small eastern states such as Delaware and Rhode Island and thinly settled states and territories from the sample.

[7]See Schaefer (1983). The harvest was higher than normal in the New South and lower than normal in the Old South.

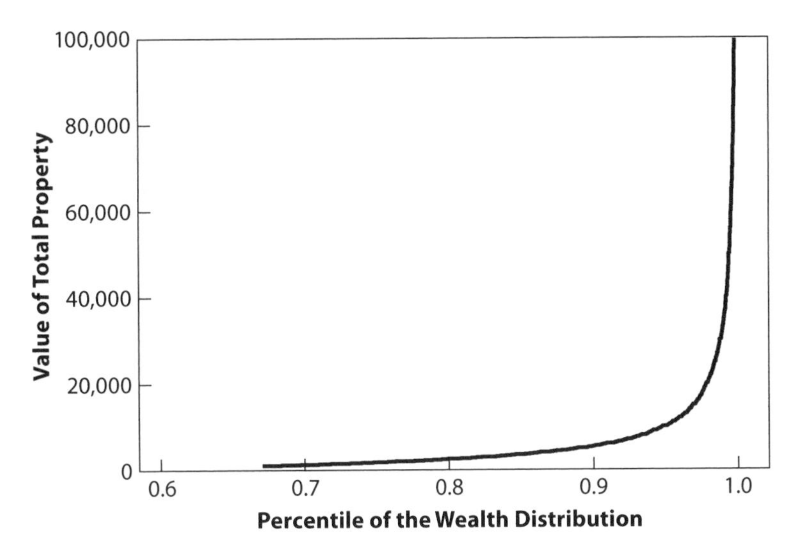

Figure 4.4. National Wealth Holdings in 1860.
Authors' calculations based on 1860 public use sample of the Census of Population.

At a national level, the elite in 1860 held 31.6 percent of the national wealth.[8] Thirty-four percent of men reported holding no real or personal property. These men tended to be young. Others held modest amounts of property. For example, 67 percent of men held $1,000 or less in total property. They held just 4.8 percent of aggregate wealth. Figure 4.4 shows the cumulative wealth distribution for the United States for adult white males. It shows the part of the distribution from $1,000 (67th percentile) to $100,000 (99.8th percentile). The distribution is quite flat up to the 90th percentile. In fact, the top 90 percent of men held only 27.4 percent of aggregate wealth. The other 10 percent held 72.6 percent of the wealth.

The amount of wealth held by the elite varied substantially across states. The elite in the top quartile states—states where the elite held the largest shares of state wealth—held 30–45 percent of the state wealth, whereas the elite in bottom quartile states held 13–19 percent of the state wealth.[9] Figure 4.5 shades

[8] This measure is computed by dividing the sum of real and personal property held by the elite by the total real and personal property in the state. In other contexts, the Gini index, the 90-10 ratio, and the 90-50 ratio may have also been used to measure wealth distributions. The problem in this context is that a very large fraction of white men report zero wealth. This makes it difficult to compute ratios and the Gini index.

[9] Table 4.1A in the appendix lists estimates of wealth by state.

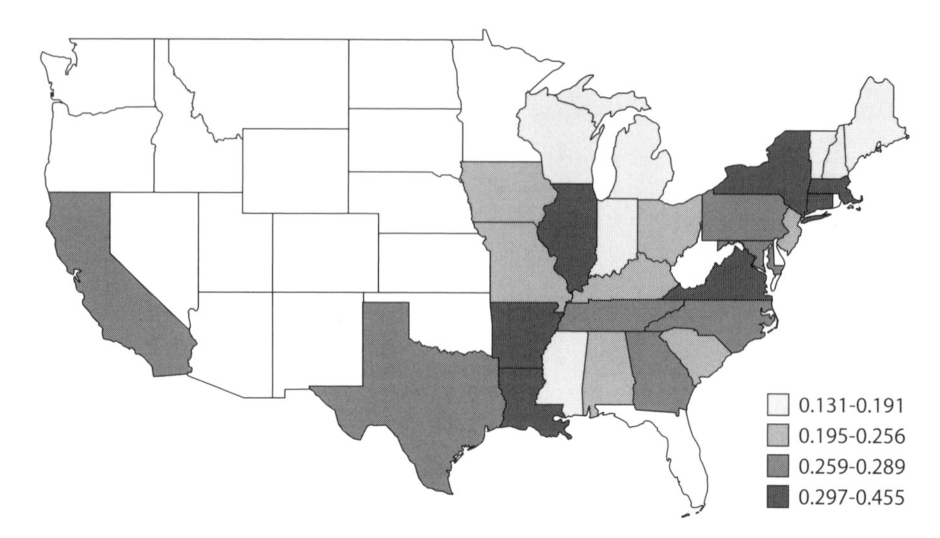

Figure 4.5. Map of Shares of State Wealth Held by the Elite in 1860.

Darker shadings indicate higher wealth holdings by the elite. White indicates that the measure could not be computed for that state. Authors' calculations based on 1860 public use sample of the Census of Population.

states' elite wealth by quartiles. Wealth was the most highly concentrated in Massachusetts, Connecticut, New York, Virginia, Illinois, Arkansas, and Louisiana. It was the least highly concentrated in Maine, Vermont, New Hampshire, Indiana, Wisconsin, Michigan, and Mississippi. Table 4.1 shows the average and the range for state wealth.

Our estimates of the share of the wealth held by the state economic elite are largely in line with Soltow's (1975) estimates. Soltow found that the top 1 percent in the North and the South each controlled 27 percent of the total assets and that the top 1 percent in the United States as a whole controlled 29 percent of the total assets. These numbers differ slightly from our estimate of 31.6 percent, because Soltow included the entire free male population over the age of twenty. Recall that the sample used here includes the white male population ages twenty-one and older. Soltow's slightly larger base will tend to depress the wealth holdings of the elite, by adding more individuals with few or no assets. Unfortunately, Soltow does not report state-level measures, and so it is impossible to compare our state-level estimates with his.[10]

[10] Our sample is four times larger than Soltow's sample. Soltow, however, oversampled persons worth more than $100,000 at forty times the rate of individuals below $100,000, so he has a larger, and possibly more accurate, sample of the very rich.

TABLE 4.1.
SUMMARY STATISTICS FOR MEASURES OF ELITES IN 1860

	Wealth share of elites	*Occupational homogeneity of elites*
Average	0.25	5,130
Maximum	0.45	10,000
Minimum	0.13	3,380

Notes: The sample includes 28 states. Authors calculations based on 1860 Census of Population public use 1–100 sample.

Our estimates are also consistent with the findings of studies of specific states and regions. For example, Pessen (1973) used tax records to construct wealth distributions for three cities—New York, Brooklyn, and Boston—during the 1840s. The top 1 percent of the wealth distributions in New York, Brooklyn, and Boston controlled 40 percent, 42 percent, and 37 percent of the noncorporate wealth. Using the Bateman-Foust sample of rural households from the 1860 Census of Population for the northern-tier states, Atack and Bateman (1981) found a "much more equal distribution [of wealth] in the rural north" than in other parts of the United States. Using tax records from townships in Massachusetts, Steckel and Moehling (2001) computed that the top 1 percent held 27 percent of the total taxable wealth in 1860.[11]

To the extent that the wealth distribution is largely persistent, the 1860 census tells us something about wealth distributions for earlier and later periods. In his book on the distribution of wealth in 1798, which made comparisons with the distribution in 1860, Soltow (1989) concluded that "There is evidence that inequality [of wealth] within states remained stable during both the eighteenth and nineteen centuries."[12]

[11] Using estate tax records from the twentieth century, Kopczuk and Saez (2004) find that the top 1 percent of all households held 40 percent of total wealth. This declined sharply in the 1930s and 1940s to 22.5 percent of total wealth in 1949.

[12] Soltow (1989), 190. Elite wealth in 1860 also predicts subsequent income distributions. Using federal tax data, Sommeiller (2006) constructed data on state-level income distributions for the period 1913–2003. In unreported regressions, states with more concentrated wealth holdings in 1860 also had more concentrated income over four twenty-year periods spanning 1920–2000. This pattern is striking, because it suggests that the structure of state wealth and income has been remarkably persistent over time, despite the many changes that occurred in state economies over the period 1860–2000.

TABLE 4.2.
DISTRIBUTION OF ELITE OCCUPATIONS IN 1860

Occupation	*Percentage*
Farmers (owners and tenants)	43.5
Managers, officials, and proprietors	28.8
Other nonoccupational response	5.2
Lawyers and judges	4.6
Physicians and surgeons	2.7
Operative and kindred workers	1.6

Notes: All occupations as coded by IPUMs for 1860 with at least 10 individuals in the elite are listed. There are 695 individuals in the elite. All individuals were white men, and all were in the elite of their own state.

Having examined the wealth shares held by state elites, we turn to their occupations. The 1860 Census of Population asked individuals about wealth and occupation. The public use sample for the 1860 Census of Population classifies similar occupations together. For example, the occupational group "farmer" includes individuals who reported that they were ranchers and plantation owners. The distribution of occupations for the state economic elites is shown in table 4.2. Nearly half of the economic elite, 44 percent, were farmers. The next most common elite group, at 29 percent, was "merchants, officials, and proprietors." Other occupational groups such as lawyers and judges (5 percent) and physicians and surgeons (3 percent) were much smaller. A catchall group, "other," was created for these smaller occupations.[13]

The occupational homogeneity of the state economic elite was computed using a Herfindahl-Hirschman index (HHI). The HHI is the sum of the squares of the occupational percentages of the state economic elite. An HHI of 10,000 would mean that all of the members of the economic elite shared the same profession. An HHI of 3,333 would mean that the members of the economic elite

[13]The results are similar if each of these smaller categories is considered separately.

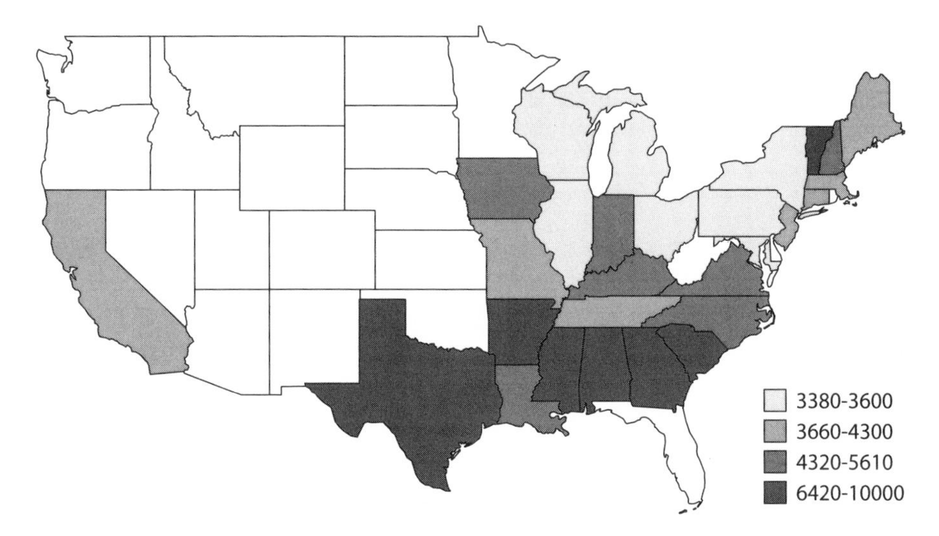

Figure 4.6. Map of Occupational Homogeneity of the Elite in 1860.

Darker shadings indicate greater occupational homogeneity of the elite. White indicates that the measure could not be computed for that state. Authors' calculations based on 1860 public use sample of the Census of Population.

were evenly divided among (1) farmers, (2) merchants, officials, and proprietors, and (3) other. In the examples in figures 4.2 and 4.3, the occupational homogeneity was $5{,}972 = [(75.00)^2 + (16.67)^2 + (8.33)^2]$.

States varied considerably in the occupational homogeneity of their elite. The elite in top-quartile states—the least diverse states—had HHIs of 6,900–10,000, whereas the elite in bottom-quartile states—the most diverse states—had HHIs of 3,400–3,700.[14] Figure 4.6 shades states' elite HHIs by quartiles. The most occupationally homogeneous elites were in Vermont, South Carolina, Georgia, Alabama, Mississippi, Arkansas, and Texas. The least occupationally homogeneous elites were in Illinois, Wisconsin, Michigan, Ohio, Pennsylvania, New York, and Maryland. Table 4.1 shows the average and range for state HHIs.

Figure 4.7 illustrates the relationship between the share of wealth of the state held by the elite and their occupational homogeneity. The two measures are relatively uncorrelated (–0.04). This low correlation is consistent with the

[14] Table 4.1A in the appendix lists estimates of occupational homogeneity by state.

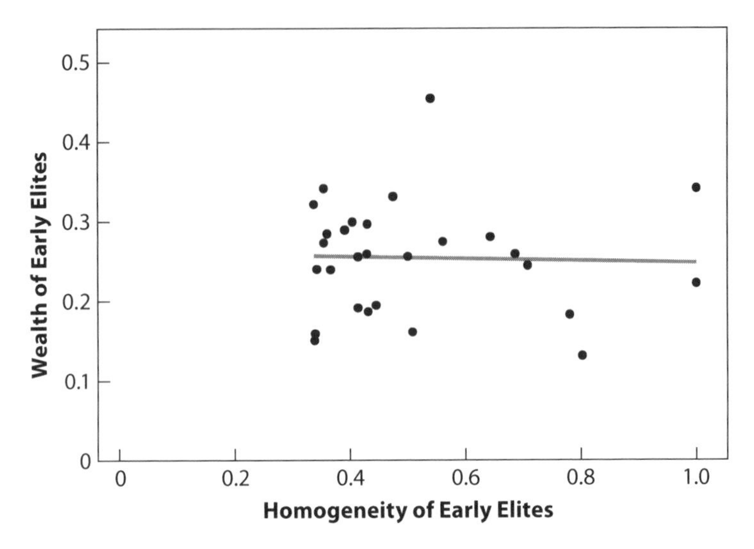

Figure 4.7. Occupational Homogeneity and Wealth of the Elite in 1860.

The R-squared for the regression of the wealth of early elites on homogeneity of early elites that includes a constant is 0.001.

patterns in figures 4.5 and 4.6. For example, the states with top quartile of wealth shares included Arkansas, Connecticut, Illinois, Louisiana, Massachusetts, New York, and Virginia. Despite having high wealth shares, they had very different levels of occupational homogeneity. Illinois and New York were in the lowest quartile for occupational homogeneity; Connecticut and Massachusetts were in the second lowest quartile; Louisiana and Virginia were in the second highest quartile; and Arkansas was in the highest quartile.

Initial Conditions, Wealth, and Occupational Homogeneity

Table 4.3 provides the correlations between the two measures of the state elite and state initial conditions. Occupational homogeneity of the elite was positively correlated with precipitation, temperature, distance to internal water, and distance to the ocean. The wealth of the elite was also correlated with these initial conditions, but the correlations are somewhat weaker, particularly for distance to the ocean. Both measures of elites are relatively uncorrelated with civil-law origins.

The fact that precipitation and temperature are positively correlated both with wealth and with occupational homogeneity is not entirely surprising. The

TABLE 4.3.
CORRELATIONS OF MEASURES OF ELITES IN 1860 AND INITIAL CONDITIONS

	Occupational homogeneity of elites	*Wealth share of elites*
Wealth share of elites	–0.036	1.000
Precipitation	0.432	0.302
Temperature	0.546	0.462
Distance to internal water	0.317	0.282
Distance to ocean	0.251	–0.013
Civil	0.208	0.116

Note: The sample includes 28 states.

South has long been recognized as having had both an unequal wealth distribution and low levels of occupational homogeneity.[15]

That a state's distance to internal water and distance to the ocean are positively related to the occupational homogeneity of its elite is less obvious. The effect of proximity to water transportation on the elite becomes evident, however, when one considers the location of the twenty largest cities in 1860.[16] Seven of the cities were ocean ports where rivers met the sea (Baltimore, Boston, Brooklyn, New York, Newark, Providence, and San Francisco); five of the cities were ports where rivers met the Great Lakes (Albany, Buffalo, Chicago, Detroit, Milwaukee, and Rochester); two of the cities were on the Mississippi River (St. Louis and New Orleans); three of the cities were on the Ohio (Cincinnati, Louisville, and Pittsburgh); two of the cities were on rivers that were close to the Atlantic (Philadelphia and Washington) and one city (Albany) was on a major river upstream from New York. These cities were located in states with easy access to multiple forms of water transportation. Much of the trade and small-scale manufacturing that occurred in the United States occurred in, or was mediated by, these cities. Many of these cities were located in New

[15] See Key (1964) and Kousser (1974).

[16] This discussion relies heavily on Glaeser and Kohlhase (2003), which discusses the proximity of major cities in 1900 to water.

York, Massachusetts, Pennsylvania, and Ohio, which produced more than half of the value of manufactured products in the United States in 1860.[17]

Being close to water would tend to diversify the economic elite away from agricultural occupations and so lower the occupational homogeneity of the elite by increasing the share of the elite in the managers, officials, and proprietors category and in the all-other category. Trade and manufacturing could also plausibly increase the share of wealth held by the elite relative to its share in other locations.

The relationships between initial conditions and the two measures of the elite are explored further in the regressions in table 4.4. The first column shows that states with greater precipitation and states that were farther from internal water had greater occupational homogeneity. The effect of precipitation was large. A 10 percent increase in precipitation increased occupational homogeneity by 8 percent. The effect of distance to internal water was somewhat smaller, but still sizeable. A 10 percent increase in distance to rivers and the Great Lakes increased occupational homogeneity by 3 percent. These patterns are consistent with states with greater precipitation or that were farther from internal water specializing in agricultural production, in part because of their endowment and in part because of limited opportunities for trade.

The second column shows that states with higher temperatures and states that were closer to oceans had elites that controlled a high fraction of the state's wealth. The large coefficient on temperature reflects the wealth held by southern elites who engaged in slave-based plantation agriculture. A 10 percent increase in temperature was associated with a 14 percent increase in wealth held by the elite. Being closer to an ocean—and thus international trade—was also associated with a higher wealth share of the elite, although the effect was small. A 10 percent decrease in distance increased elite wealth share by 1 percent. Thus, initial conditions were associated with the characteristics of the state elite. Moreover, the patterns were consistent with their influence occurring through agriculture and trade.

The Elite and State Politics

Although the foregoing analysis has focused on the elite, one question is whether this is the correct focus. Focusing on the elite is likely to be correct

[17] Pessen (1973), in his work on antebellum wealth in the Northeast, focused on New York, Brooklyn, Boston, and Philadelphia, because the richest men lived in those cities.

Table 4.4.
Initial Conditions and Measures of Elites in 1860

Dependent variable	*Log of occupational homogeneity, 1860*	*Log of wealth share, 1860*
Log of precipitation	0.799**	0.0173
	(0.349)	(0.304)
Log of temperature	–0.0564	1.404***
	(0.801)	(0.446)
Log of distance to internal water	0.260**	–0.0367
	(0.109)	(0.108)
Log of distance to ocean	0.0936	–0.0987*
	(0.0830)	(0.0539)
Civil	–0.0306	–0.0595
	(0.128)	(0.138)
Constant	5.558**	–1.647
	(2.259)	(1.405)
Observations	28	28
R^2	0.478	0.335

Notes: Standard errors are heteroskedasticity robust. The notations *, **, and *** denote significance at the 10 percent, 5 percent, and 1 percent levels.

if for no other reason than that many early state constitutions had property requirements to hold office.[18]

The historical evidence suggests that nineteenth-century state legislators were wealthy. Some were members of the economic elite. Wooster's outstanding books (1969, 1975) on the Upper and Lower South provide detailed evidence

[18] Greene and Pole (2000), 279. It is worth noting that a strand of the political history literature has focused on voters—their participation in the political process, their identification with parties, and their reaction to policy outcomes. Formisano (1994, 2001) reviews this literature. One problem with these studies is that "Private motives of elites, as well as long-range patterns of social and economic development, are ignored as sources of economic policy." McCormick (1974), 375.

on the wealth of state legislators in 1860. Table 4.5 shows that the median state legislator in the Upper and Lower South held substantially more assets than the 90th percentile of the wealth distribution. With few exceptions, the median wealth of members of the state house fell between the 90th and the 95th percentile, and the median wealth of members of the state senate fell between the 95th and the 99th percentile. One reason that these men fell below the 99th percentile is that many were in their early forties and so had not yet finished accumulating assets. Some of these men would go on to be the economic elite or had fathers or brothers in the elite. Others would be cultivated by the elites, as one did not have to be in the legislature to have influence.

Other evidence also suggests that elected officials were wealthy.[19] Unfortunately, studies rarely offer the level of detail of Wooster's work or cover more than one city or state. In a review of the available historical evidence on the characteristics of officeholders, Pessen (1980) concluded:

> The resultant picture inevitably is not uniform. Humble county and town officials, for example, were less likely to be drawn from the highest levels of wealth and from the most prestigious occupations than were men who occupied more exalted state and federal positions. Alderman and councilmen usually did not match the mayor either in wealth or in family prestige. But the relatively slight social and economic differences found between men at different levels of government or between men nominated by the parties that dominated American politics from the 1830s to the 1850s were not differences between the North and the South. In the South as in the North, men similar in their dissimilarity to their constituencies held office and exercised behind-the-scenes influence. In contrast to the small farmers, indigents, laborers, artisans, clerks, and shopkeepers—the men of little or no property who constituted the great majority of the antebellum population—the men who held office and controlled the affairs of the major parties were everywhere lawyers, merchants, businessmen, and relatively large property owners.[20]

Low pay meant that only the wealthy were likely to serve as elected officials. Offices were a source of prestige. Depending on the position, officeholders may have earned additional income from graft, investment opportunities, or other sources. But in the majority of cases, officeholders had to have family wealth or a primary occupation that provided both flexibility and income. For example, in most states, legislators would have needed considerable outside

[19] See Watson (1997).
[20] Pessen (1980), 1137–1138.

Table 4.5.
Wealth of State Legislators in the Upper and Lower South in 1860

State	*Median wealth of legislator in 1860*	*90th percentile of wealth distribution*	*95th percentile of wealth distribution*	*99th percentile of wealth distribution*
Alabama	21,000 (H) 58,500 (S)	13,370	27,000	86,000
Arkansas	9,000 (H) 18,000 (S)	6,000	13,900	80,000
Florida	9,000 (H) 52,000 (S)	7,400	14,500	44,000
Georgia	13,000 (H) 21,000 (S)	8,500	18,360	62,000
Kentucky	9,250 (H) 12,000 (S)	6,000	11,010	38,000
Louisiana	18,000 (H) 35,839 (S)	10,000	25,000	191,130
Maryland	11,250 (H) 33,150 (S)	6,000	14,000	40,870
Mississippi	22,000 (H) 27,500 (S)	19,270	37,000	103,000
Missouri	8,300 (H) NA (S)	4,750	8,460	30,000
North Carolina	17,000 (H) 31,000 (S)	6,800	16,000	54,300
South Carolina	32,000 (H) 70,000 (S)	20,000	33,300	110,000
Tennessee	14,000 (H) 11,500 (S)	8,000	16,030	50,960
Texas	18,600 (H) 25,000 (S)	9,600	16,630	62,000
Virginia	17,000 (H) 35,000 (S)	11,460	21,500	72,600

Sources: Data for the Upper South are from Wooster (1975), table 6, p. 35, and table 8, p. 38. Data for the Lower South are from Wooster (1969) Table 4 (p. 39) and Table 5 (p. 40). Percentiles are authors' calculations based on 1860 public use sample of the Census of Population.

income to reach the earnings of manufacturing workers. Members of Congress met for 210 days and made $7,500 in 1910.[21] At that time, state legislators met for 28 days and made between $100 and $1,500 in salary and per diem. New York, at $1500, was an outlier. The median value was less than $200. In comparison, the average annual earnings for manufacturing workers in 1909 were $512.[22] Most other officials, with the exception of judges and the governor, made less than legislators.

Not only were legislators and other officials wealthy, their selection was influenced by the economic elite. In his discussion of the problems of representation in state government, Harvey (1949) noted that "Property interests dominated American state governments during the eighteenth and nineteenth centuries. With the disappearance of property qualifications for holding a seat in the legislature and for voting, these interests nevertheless did not go into eclipse. They supported party organization by contributing heavily to campaign funds; and in return they demanded, and usually received, certain favors or control over candidates elected."[23]

How was the occupational homogeneity of the elite likely to affect politics? Occupational homogeneity is expected to directly affect political competition, because homogeneous elites will tend to support a single party, whereas heterogeneous elites will tend to support different parties.

Historical evidence suggests that elite occupations played a role in politics. In his study of revolutionary Philadelphia, Doerflinger (1986) writes:

> The destruction of traditional [British] political elites, the upsurge in popular political participation, and the emergence of divisive economic issues during the war had eroded the values of mixed government and converted occupational groups into organized, articulate political factions. . . . The recasting of political participation along occupational lines was remarked on by contemporaries and seemed to be a fundamental trait of modern republics. As James Madison observed in Federalist 10, "The most common and durable source of factions has been the various and unequal distribution of property. . . . A landed interest, a manufacturing interest, a mercantile interest, a moneyed interest, with many lesser interests, grow up of necessity in civilized nations, and divide them into different classes, actuated by different sentiments and views."[24]

[21] Squire and Hamm (2005), 72.

[22] Rees (1975), 32.

[23] Harvey (1949), 267.

[24] Doerflinger (1986), 276. See also Benson (1955, 1960, 1961), Campbell (1980), Wilentz (1982).

Both Dalzell (1987) and Pessen (1973) discuss the political activities of the elite merchants they study. These merchants on average tended to be Whigs. Goodman (1986) writes, "Central to any understanding of Rhode Island politics in the Jacksonian era was polarization between the northern industrial towns, with Providence at the center, which favored the Whigs, and the rural towns in southern Rhode Island, which favored the Democrats and had dominated the state owing to an antiquated colonial charter that favored the landholders."[25] Thus, political competition within the elite ran along occupational lines in some times and places.[26]

Contemporary evidence also suggests that occupation plays a role in party affiliation.[27] For example, Day and Hadley (2001) find important occupational differences among donors to Democratic and Republican political action committees devoted to the election of women. Hout, Brooks, and Manza (1995) examine the voting behavior of six occupational groups over the period 1948–1992. They find that individuals with similar occupations tend to have similar political affiliations.

Wealth, Occupational Homogeneity, and State Political Competition

Having examined historical evidence on the elite and politics, we turn to a central question: How are measures of the elite in 1860 related to political competition?

The focus has been on occupational homogeneity as the mechanism through which initial conditions acted on political institutions and more specifically state legislatures. Two other mechanisms that link initial conditions and institutions are also relevant. One mechanism, due to Engerman and Sokoloff (1997, 2000), focuses on how soil and climate at time of settlement shaped elite wealth and institutional outcomes. In countries where the soil and climate were conducive to slave-based production of goods such as sugar and cotton, elites amassed substantial wealth, and institutions that narrowly represented elite interests emerged. Where soil and climate were not conducive to slave-based production, elites accumulated relatively less wealth, and institutions

[25] Goodman (1986), 44.

[26] Formisano (1994) also writes: "Economic and political elites at the local, regional, or national level were not always united and self-conscious about their goals, but on balance they were distinctively more conscious and cohesive in pursuit of their goals than artisans, workers, and laborers" (474).

[27] Although most of the emphasis in the voting literature has been on income to the exclusion of occupation, a few studies examine occupation.

developed that represented broader public interests. A second mechanism, due to Acemoglu, Johnson, and Robinson (2001), focuses on how the disease environment at time of settlement shaped institutions. When early settlers of European colonies were at high risk of being stricken with fatal diseases, they tended to quickly extract wealth. When these settlers faced lower-risk disease environments, they invested in longer-term projects. Thus, institutions that allowed expropriation of labor and wealth emerged in colonies that had high mortality among early settlers, and institutions that protected property rights emerged in colonies that had low mortality among early settlers.

What these mechanisms have to say in the context of the American states is somewhat unclear, because their discussion focuses on political institutions and not on political competition. The first mechanism is plausibly relevant for political competition. States with very wealthy early elites would have weak political competition over long periods. States with less wealthy early elites would either always have high political competition or would start with weak political competition but evolve towards stronger political competition as the elite's ability to maintain control waned. The relevance of the second mechanism is less obvious, because our context differs substantially from the colonial context that Acemoglu, Johnson, and Robinson (2001) study. If, despite these differences, elite incentives differed with mortality, then southern states would have had lower levels of state political competition.

Figures 4.8–4.14 document the negative relationship between occupational homogeneity of the elite and political competition over the period 1866–2000. The relationships between measures of the state elite and average political competition in the state legislature are plotted at twenty-year intervals. Taken together, the figures suggest the occupational composition of elites had a persistent influence on politics during 1866–2000.

In contrast, the relationship between elite wealth and political competition was not of the predicted sign for most of the period. Through the end of the 1930s, the relationship between elite wealth and political competition was positive or zero. During the 1940s and 1950s, the relationship was only slightly negative, and beginning in the 1960s it was strongly negative. The initially positive relationship between wealth and political competition is surprising, since greater elite wealth is generally thought to lead to lower-quality institutions and lower levels of political competition.

These general patterns are confirmed in aggregate and in the repeated cross-sections in table 4.6, which control for temperature, precipitation, distance to

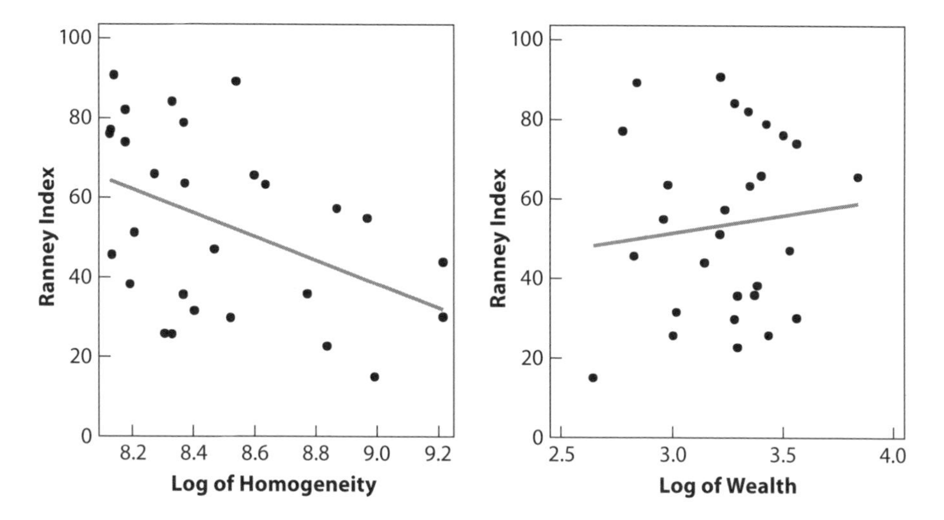

Figure 4.8. State Political Competition and State Elites, 1866–1878.

In figures 4.8–4.14, log of homogeneity equals the log of 1 + HHI, and the log of wealth is the log of (1 + share wealth of early elites). Both occupational homogeneity of the elite and wealth of the elite are measured in 1860. The *R*-squares for the regression of the Ranney in 1866–1878 on log of homogeneity and then on log of wealth are 0.191 and 0.012.

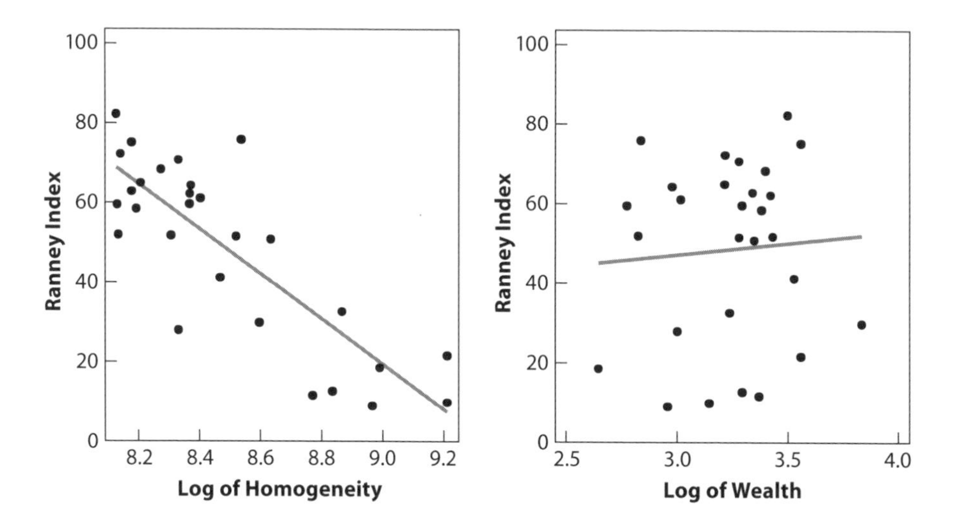

Figure 4.9. State Political Competition and State Elites, 1880–1898.

The *R*-squares for the regression of the Ranney in 1888–1898 on log of homogeneity and then on log of wealth are 0.681 and 0.005.

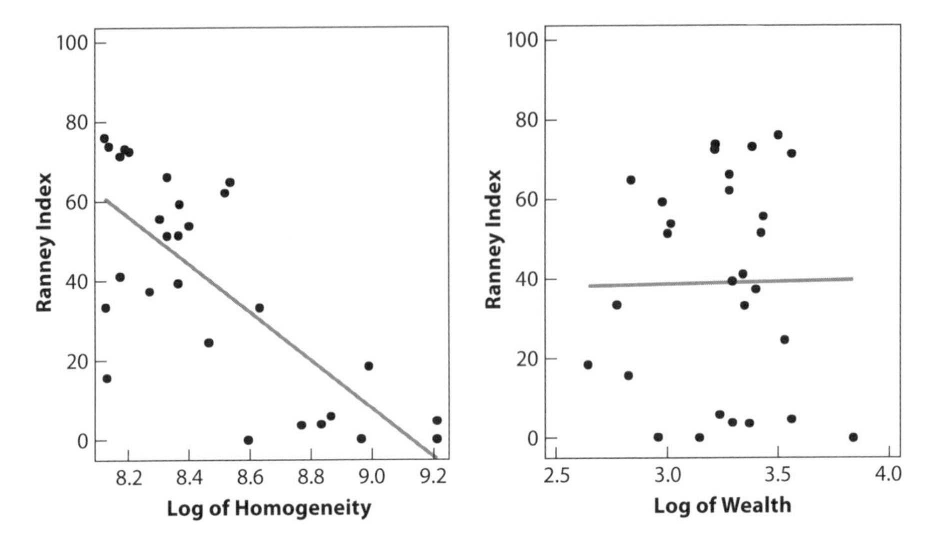

Figure 4.10. State Political Competition and State Elites, 1900–1918.

The *R*-squares for the regression of the Ranney in 1900–1918 on log of homogeneity and then on log of wealth are 0.552 and 0.000.

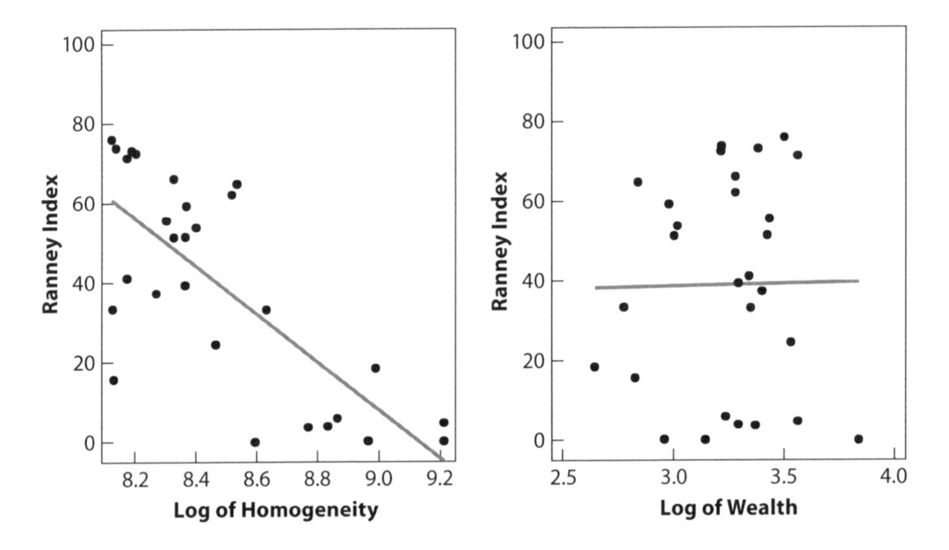

Figure 4.11. State Political Competition and State Elites, 1920–1938.

The *R*-squares for the regression of the Ranney in 1920–1938 on log of homogeneity and then on log of wealth are 0.516 and 0.000.

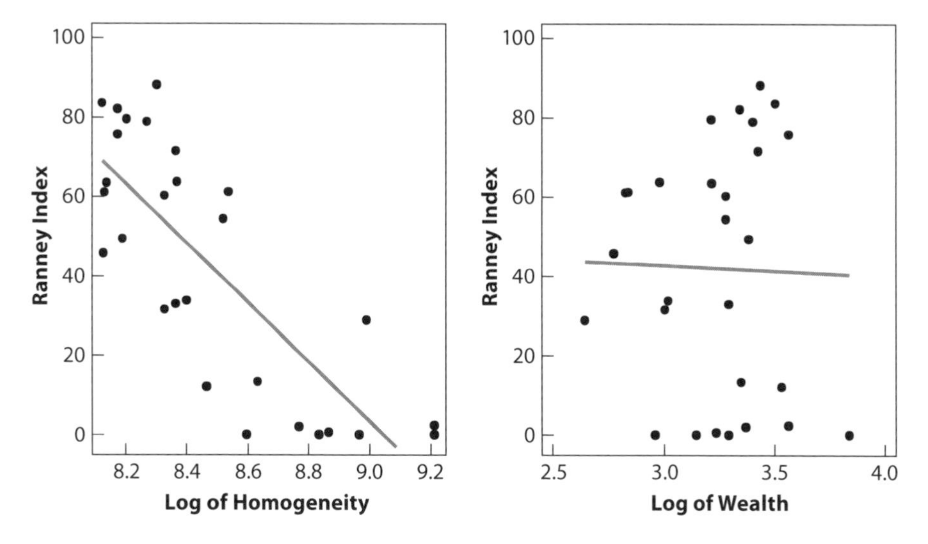

Figure 4.12. State Political Competition and State Elites, 1940–1958.

The *R*-squares for the regression of the Ranney in 1940–1958 on log of homogeneity and then on log of wealth are 0.629 and 0.000.

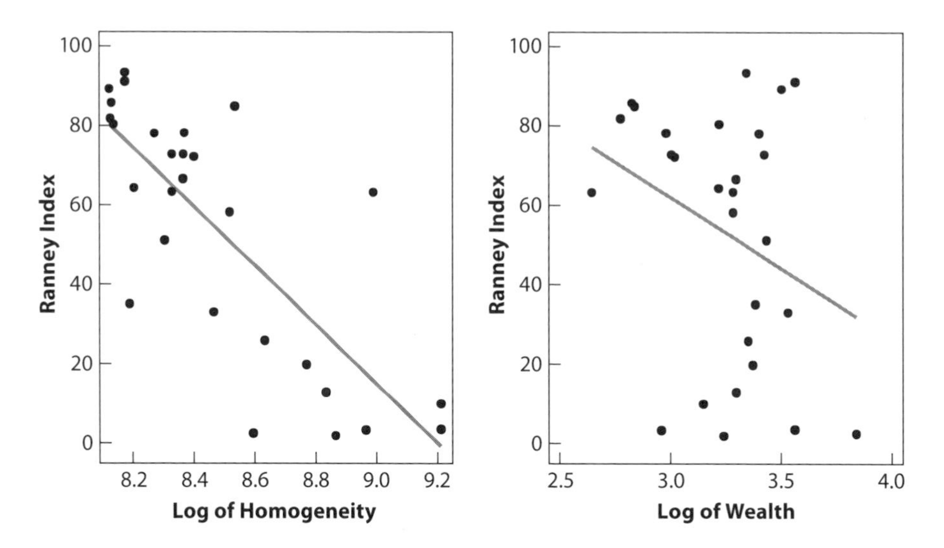

Figure 4.13. State Political Competition and State Elites, 1960–1978.

The *R*-squares for the regression of the Ranney in 1960–1978 on log of homogeneity and then on log of wealth are 0.593 and 0.062.

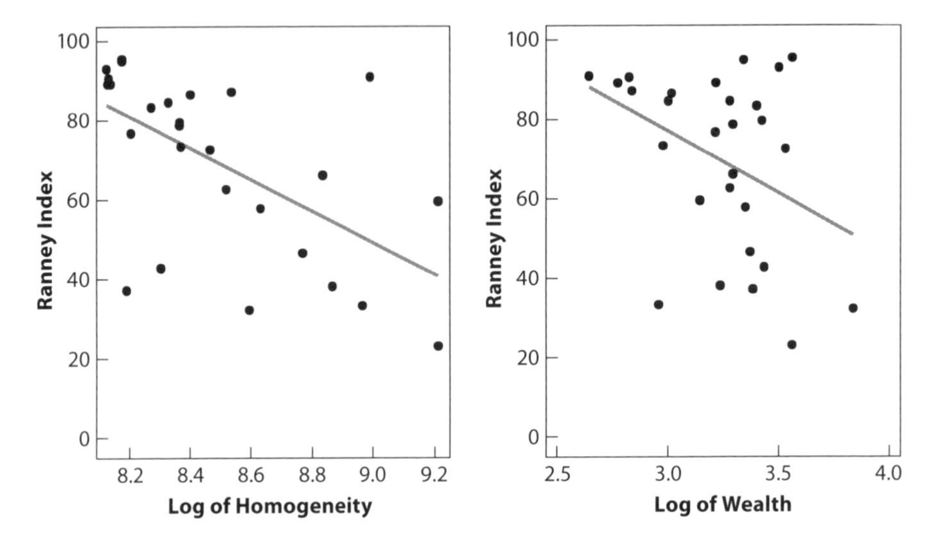

Figure 4.14. State Political Competition and State Elites, 1980–2000.

The R-squares for the regression of the Ranney in 1980–2000 on log of homogeneity and then on log of wealth are 0.321 and 0.116.

internal water, and distance to oceans.[28] The regressions in the top panel of table 4.6 demonstrate that the relationships in figure 4.1 are empirically relevant. The first column shows that over the period 1866–2000, occupational homogeneity of the elite in 1860 was strongly negatively related to state political competition, even when controls were included for state initial conditions. A 10 percent increase in homogeneity was associated with a four-point drop in the state Ranney index. The average state Ranney index over this period was 49, so a four-point drop was sizeable. Further, in six of the seven periods, the relationship between occupational homogeneity and political competition was negative and statistically significant. States with high occupational homogeneity in 1860 had lower levels of state political competition during all of the shorter periods except 1980–2000. The next section argues that the relationship between occupational homogeneity of the elite and state political competition was causal. We defer discussion of the coefficients on the initial conditions until then.

[28]The panel analogue to the repeated cross section regressions is presented in appendix table 4.2A. Given the civil laws' lack of theoretical and empirical importance in chapter 3 and in table 4.4, it is not included here or in later regressions.

The bottom panel of table 4.6 investigates the relationships between wealth share of the elite in 1860 and political competition when controls were included for initial conditions. The first column shows that over the period 1866–2000 wealth share of the elite in 1860 was strongly positively related to state political competition. A 10 percent increase in wealth share was associated with a three-point increase in the state Ranney index—which is surprising given that the literature has suggested that wealth concentration is negatively related to political competition. In the shorter periods, wealth share was always positively related to political competition. In four of the seven periods, the relationship between wealth and political competition was statistically significant. States with a higher share of wealth held by the elites had higher political competition during periods spanning 1880–1958.

Given the limited theoretical grounding for the importance of wealth in the United States and its positive sign, which is contrary to the international literature, the remaining discussion focuses on occupational homogeneity of the elite.

Causality

Our mechanism offers a causal role to the elite, but the statistical evidence thus far only shows a correlation between the occupational homogeneity of the state elite and state political competition. To make a statement about causality, one needs to address possible endogeneity by using an instrument. Below we describe an instrument for occupational homogeneity that arguably satisfies the exclusion restriction and present results using the instrument. The major constraint on the analysis that follows is the fact that data are only available for twenty-eight states.

The most obvious potential instruments are the initial conditions. Measures of climate arguably fail to satisfy the exclusion restriction. That is, their influence in likely to continue to be important because of the importance of agriculture as an occupation in most states well into the twentieth century. Further, temperature and precipitation are strongly correlated with the South, which had persistently lower political competition throughout most of the twentieth century.

Distance to internal water both played a role in shaping the elite and is likely to satisfy the exclusion restriction, because of the rise of the railroad. How did the introduction of the railroads affect the importance of internal water

Table 4.6.

Initial Conditions, Early Elites, and State Ranney Index

Dependent Variable	*Ranney 1866–2000*	*Ranney 1866–1878*	*Ranney 1880–1898*	*Ranney 1900–1918*	*Ranney 1920–1938*	*Ranney 1940–1958*	*Ranney 1960–1978*	*Ranney 1980–2000*
Log of occ. homogeneity	–43.61***	–23.78**	–53.65***	–57.40***	–50.50***	–58.38***	–42.68***	–11.16
	(6.535)	(10.33)	(6.889)	(9.719)	(8.545)	(12.59)	(9.764)	(13.05)
Log of precipitation	–16.55	–21.77	–10.31	22.41	14.28	–22.56	–36.96	–57.27***
	(20.38)	(27.80)	(24.60)	(27.54)	(26.80)	(34.23)	(24.54)	(16.25)
Log of temperature	–23.89	132.6***	28.21	–32.68	–47.81	–31.23	–99.09***	–44.83
	(25.15)	(40.36)	(38.95)	(29.31)	(43.52)	(53.24)	(25.10)	(31.70)
Log of distance to internal water	–9.030	–26.71**	–11.23	–6.178	–3.423	–5.142	–7.135	–10.39
	(7.969)	(9.515)	(9.489)	(9.643)	(11.69)	(13.75)	(8.619)	(6.429)
Log of distance to ocean	3.817	–8.853*	3.587	6.638	6.207	0.487	7.393*	4.898
	(3.207)	(4.887)	(2.928)	(5.358)	(4.323)	(5.647)	(4.036)	(4.852)

R^2	0.787	0.437	0.728	0.628	0.610	0.673	0.845	0.665
Log of wealth share	27.87**	5.785	35.41**	40.68**	41.57**	43.68**	19.69	0.134
	(12.41)	(15.94)	(13.48)	(17.86)	(16.31)	(17.29)	(12.89)	(11.22)
Log of precipitation	−54.27**	−41.73	−56.79*	−27.50	−30.01	−73.47*	−73.38**	−66.47***
	(25.16)	(29.99)	(31.13)	(26.66)	(25.81)	(36.78)	(31.36)	(15.90)
Log of temperature	−57.81	126.8**	−15.05	−82.74	−99.66*	−85.25	−122.1***	−44.07
	(36.98)	(50.23)	(49.41)	(54.04)	(50.39)	(57.04)	(37.86)	(38.09)
Log of distance to internal water	−18.75**	−32.47***	−23.14**	−18.78	−14.23	−17.85	−17.02*	−13.22*
	(8.083)	(9.854)	(9.341)	(10.97)	(10.62)	(12.19)	(9.077)	(6.699)
Log of distance to ocean	2.825	−10.39*	2.484	5.753	6.038	−0.169	5.622	3.906
	(5.147)	(5.117)	(5.009)	(7.598)	(6.245)	(7.300)	(5.451)	(5.484)
R^2	0.657	0.377	0.532	0.485	0.522	0.574	0.761	0.651

Notes: The notations *, **, and *** denote significance at the 10 percent, 5 percent, and 1 percent levels. Standard errors are robust. There are 28 observations in each estimate. Constant is estimated but not reported.

transportation? Many places that previously had shipped goods long distances to water sources or had not shipped goods at all were suddenly able to move goods at relatively low cost on the railroad.

Use of rivers such as the Mississippi and its tributaries shifted markedly following the rise of the railroad. Fishlow (1964) writes, "Laments for the decline of New Orleans, as a site of western receipts did not blame declining southern appetites, but, properly, focused on the rapid construction of rail feeders that narrowed the economic hinterland of New Orleans. Nowhere was the shift more obvious than in the Ohio Valley. The proportion of flour flowing eastward or northward from Cincinnati increased from 3 percent in the early 1850s to 90 percent in 1860; similarly for pork, there was a shift from 7 percent to 42 percent."[29]

By 1890 railroad coverage was dense in most areas.[30] Railroads had opened up hinterlands. Indeed, more than half of the direct benefits of the railroad calculated by Fishlow (1964) came from opening up of the hinterlands. Fogel (1964) computed that in 1890 in the absence of the railroad, shipping agricultural commodities interregionally would have been substantially more expensive. "Since the actual 1890 cost of shipping the specified commodities was approximately $88,000,000, the absence of the railroad would have almost doubled the cost of shipping agricultural commodities interregionally. It is therefore quite easy to see why the great bulk of agricultural commodities was actually sent to the East by rail, with water transportation used only over a few favorable routes."[31]

The rise of railroads had plausibly broken the link between distance to rivers and lakes and political competition. The decline in importance of internal water transportation was only reinforced with later introduction of road and air transport. Thus, distance to rivers and lakes may be a suitable instrument.

Before it can be used as an instrument, one concern needs to be addressed. If railroads were merely replicating existing water transportation, then distance to rivers and Great Lakes would not be useful as an instrument. In general, states with shorter distances to rivers and Great Lakes also had more railroads. The railroad, however, went beyond existing water networks. Atack et al. (2009) found that for Midwestern counties, whether a county gained access to a railroad was strongly negatively related to whether the county had access to a navigable river and unrelated to whether a county had access to a Great Lake.

[29] Fishlow (1964), 335.

[30] It is worth noting that access to water changed very little after 1850. See Atack et al. (2009), 13.

[31] Fogel (1964), 211.

Together with the evidence from Fogel (1964) and Fishlow (1964), this suggests that the railroad was not simply a substitute for water transportation—it was opening up new hinterlands. And in these locations, it was dramatically reducing the effective distance to markets.

Distance to oceans played a lesser role in shaping the elite and is unlikely to have satisfied the exclusion restriction. Having a railroad might increase the ability of merchants in towns in Kansas to engage in internal trade, but their ability to broker international trade was still limited by location. That is, merchants in major port cities like New York almost surely still had a comparative advantage in international trade. As a result, it is more difficult to argue that the rise of the railroad somehow broke the relationship between distance to oceans and state political competition. As in figure 4.1, we take a conservative approach that allows distance to oceans to have an independent effect on state political competition.

Table 4.7 presents the instrumental variables estimates of the effect of the occupational homogeneity of the elite in 1860 on state political competition. The results are presented without and with California. Figure 4.5 showed the states for which occupational homogeneity could be computed in 1860. California is an outlier, in the sense that it is very distant from all of the other states. Because of its location, California had by far the largest average distance to navigable rivers and the Great Lakes. The fact that California is such an outlier weakens the first-stage relationship between distance to internal water and occupational homogeneity of the elite.

In the top panel, which excludes California, the occupational homogeneity of the elite has a large negative and statistically significant effect on state political competition. Having a very homogeneous elite in 1860 lowered state political competition over the period 1866–2000. For example, having a state elite that was 10 percent more homogeneous lowered the state Ranney index by nine points over the period 1866–2000 and by amounts ranging from five to thirteen points over the intervals spanning 1866–2000. These numbers are large relative to estimates in table 4.6, which were four points over the period 1866–2000 and ranged from one to six points over this interval. They are also large relative to the average Ranney index for the period, which was 49.

Interestingly, the other initial conditions had insignificant coefficients over the period 1866–2000 and had statistically significant effects in just a few of the shorter time periods. Precipitation was statistically significant in 1880–1898, but the point estimate was very close to the point estimate in other years. Both temperature and distance to ocean were statistically significant in 1960–1978. This was a period of exceptional political turmoil.

Table 4.7.
Two-Stage least Squares Estimates of Occupational Homogeneity of Early Elites on State Ranney Index

	Ranney	*Panel A: California excluded*						
Dependent variable	*1866–2000*	*1866–1878*	*1880–1898*	*1900–1918*	*1920–1938*	*1940–1958*	*1960–1978*	*1980–2000*
Log of occupational homogeneity	–93.97***	–130.2***	–113.1***	–90.85***	–80.86**	–110.2**	–95.79***	–51.90**
	(25.54)	(41.00)	(33.45)	(27.57)	(36.60)	(40.22)	(24.25)	(19.54)
Log of precipitation	37.86	67.05	51.98**	57.92	54.01	47.59	28.61	–23.85
	(27.01)	(46.44)	(23.97)	(38.70)	(33.95)	(35.22)	(24.12)	(20.80)
Log of temperature	–27.88	123.7	23.46	–35.35	–50.08	–35.05	–103.1*	–48.27
	(48.50)	(82.82)	(59.32)	(35.97)	(49.12)	(68.06)	(50.74)	(49.29)
Log of distance to ocean	9.42	0.841	10.04	10.31	10.15	7.406	13.97**	8.561
	(5.86)	(12.12)	(6.864)	(7.116)	(6.179)	(7.664)	(6.206)	(6.011)

	Ranney	*Panel B: California included*						
	1866–2000	*1866–1878*	*1880–1898*	*1900–1918*	*1920–1938*	*1940–1958*	*1960–1978*	*1980–2000*
Log of occ. homogeneity	−79.14**	−128.9**	−97.85**	−81.71*	−63.97	−78.61	−70.75*	−52.05*
	[−445.96, −38.5]	[−1,164.5, −57.9]	[−548.3, −53.0]	[−385.1, −1.5]	[−284.7, 33.2]	[−361.8, 18.9]	[−380.5, −18.8]	[−477.0, −2.4]
Log of precipitation	12.71	64.76**	26.08	42.43**	25.37	−5.903	−13.85	−23.61
	(26.16)	(29.34)	(23.24)	(20.29)	(25.25)	(37.83)	(28.14)	(15.13)
Log of temperature	−26.88	123.7	24.48	−34.74	−48.95	−32.93	−101.5***	−48.28
	(35.54)	(80.63)	(51.53)	(31.11)	(43.07)	(55.66)	(35.88)	(48.53)
Log of distance to ocean	7.02	0.623	7.572	8.830	7.421	2.312	9.925*	8.584*
	(4.78)	(10.00)	(5.629)	(5.738)	(5.832)	(7.169)	(5.024)	(4.935)

Notes on top panel: All cells contains point estimates and heteroskedasticity robust standard errors in parentheses. The notations *, **, and *** denote significance at the 10 percent, 5 percent, and 1 percent levels. Log of distance to internal water is a reasonably strong instrument because it has a first stage F-statistic of 11.1. There are 27 observations and the constant is estimated but not reported.

Notes on bottom panel: Including California weakens the log of distance to internal water as an instrument (it has a first stage F-statistic of 6.9). To make reliable inferences about the impact of the log of HHI on early elites, the 90 percent conditional likelihood confidence interval is reported (see Moreira 2003) in brackets under the standard error. The notations *, **, and *** denote significance at the 10 percent, 5 percent, and 1 percent levels.There are 28 observations and the constant is estimated but not reported.

In the bottom panel of table 4.7, which includes California, the magnitudes of the coefficients on occupational homogeneity are somewhat smaller and the levels of statistical significance are generally lower. Having a state elite that was 10 percent more homogeneous lowered the state Ranney index by eight instead of nine points over the period 1866–2000. In two periods, 1920–1938 and 1940–1958, the coefficients are no longer statistically significant. Although the point estimates for the beginning and ending periods are nearly identical, the point estimates in the intervening years are up to 30 percent lower than the estimates without California.

In sum, the evidence suggests that occupational homogeneity of the elite in 1860 had a large negative causal effect on the subsequent evolution of political competition. The next section describes the mechanism through which the early elite were able to have such an enduring effect on political competition.

Persistence

The previous discussion has focused on the early elite, but the underlying reasons for persistence warrant further discussion. Persistence occurred as the result of economic and political factors. On the economic side, the primary economic activities in geographic units changed slowly. Consequently, the elites and the party affiliation of legislators were likely to be the same over time. Changes in the elites could occur because of changes in the relative payoff to different occupations. Urbanization in a particular area might eventually lead to the decline of traditional agricultural elites. Party platforms might change, causing some elites to switch. And exogenous shocks such as the Civil War could cause changes in elite preferences as well. To the extent that change was the result of similar events in all states, relative levels of political competition would remain similar over time.

Local elites almost surely took actions that helped them maintain political power in the face of a changing population of voters and a changing local economy. Acemoglu and Robinson (2008) present a model in which changes in political institutions create incentives for elites to invest in de facto political power. They show that in some cases, this can partially or fully offset the changes in de jure political power. McCormick (1979) discussed elite response to the events of the late nineteenth century. "Southern blacks and poor whites, by participating in the Populist movement, and new immigrants, by supporting the most corrupt city machines and flirting with socialism, convinced elites everywhere that unlimited suffrage fueled disorder. Under the banner of 'reform,' they enacted registration requirements, ballot laws, and other measures to restrict suffrage and reduce the discipline of party machines."[32]

At the state level, persistence was also helped along by politics. Prior to the 1960s, state legislatures were malapportioned: "many representative districts had not been altered for decades despite a growing and mobile population."[33] Following the Supreme Court rulings in *Baker v. Carr* (1962) and *Reynolds v. Simms* (1964), forty-six states had to redistrict.[34] The need for redistricting is not entirely surprising. Reapportionment was costly and could disrupt the existing power structure, so legislatures often declined to reapportion even when required to do so by the state constitution.[35] State courts were generally unwilling to intervene. Despite the constitutional requirement, they considered apportionment a political matter.[36]

The problem of malapportionment was not new.[37] Constitutional scholar G. Alan Tarr (1996) writes:

> The states' failure to deal successfully with legislative apportionment during the first half of the nineteenth century meant that the issue continued to fester, provoking periodical efforts to remedy inequitities. Although at times these efforts succeeded, often the resistance of rural legislators increased with the disparities in population among legislative districts. For if the population disparities were substantial, remedying them would entail a major shift in political power. Beginning in the late nineteenth century, rural legislators found allies among Republican reformers, who claimed that republican government required that rural communities not be overwhelmed by the legislative power of urban areas dominated by (Democratic) party machines. Often these coalitions made reapportionment impossible.[38]

Even when reapportionment did occur, inequities tended to be only partially remedied. This partial solution tended to reflect the political bargains necessary to get any reapportionment through the state legislature.[39]

[32] McCormick (1979), 295.
[33] Sturm (1982), 72.
[34] See Tyler (1962).
[35] Tarr (2000), 146.
[36] Harvey (1949), 270.
[37] See Gardner (2006) for a discussion of "the many other forms of opportunistic manipulation of electoral rules to which politically dominant groups have resorted to entrench their dominance." Gardner (2006), 894.
[38] Tarr (1996), 6.
[39] Tarr (2000), 102–105; Dunbar and May (1995), 548–551; Hardy et al. (1981).

Together, economics and politics led to enormous persistence in the party affiliation of legislators from geographic units. This translated into enormous persistence in the relative levels of political competition across states.

THE CIVIL WAR

We have deferred discussion of slavery and the Civil War so that the mechanism and causality could be addressed. The question remains: Is the persistent negative relationship between occupational homogeneity of elites in 1860 and political competition from 1866–1978 just an artifact of the Civil War? This section presents some evidence that the answer is no—the relationship appears not to have just been an artifact of the Civil War.

There is reason to believe that our results on occupational homogeneity will survive the inclusion of slavery. To some degree, slavery and the Civil War were implicitly being controlled for through temperature and precipitation. Further, the measures of the elite (marginally) predate the Civil War. At the same time, they were almost certainly influenced by slavery.

Occupational homogeneity was not exclusively a southern phenomenon. Figure 4.6 showed that Vermont was very occupationally homogeneous, and Indiana, Iowa, Kentucky, and New Hampshire were all quite homogeneous. Figure 4.15 presents box plots of occupational homogeneity of the elite in 1860 in the North and the South. The distributions do overlap. Tennessee had the least homogeneous elite of the ten southern states in our sample. Six—the five mentioned above and Connecticut—of the eighteen northern states had occupational homogeneity that exceeded the levels in Tennessee. Vermont had the most homogeneous elite of the northern states. Eight of the ten southern states were less occupationally homogeneous than Vermont.

Table 4.8 replicates our two-stage least squares estimates from table 4.7, but includes the percentage of the state population that were slaves in 1860 as an additional control variable. If the importance of occupational homogeneity is solely coming from its correlation with slavery and the Civil War, then controlling for slavery in these regressions should render the coefficient on occupational homogeneity small and not statistically significant. Although the inclusion of slavery does diminish the magnitude of the coefficient on occupational homogeneity of the elite for the period 1866–2000 from –94 to –78 in the top panel, the effect of occupational homogeneity on political competition remains large, negative, and statistically significant. Even controlling for slavery, states that were 10 percent more homogeneous had Ranney indices that were eight points lower. The results in the remaining columns in the top panel show that

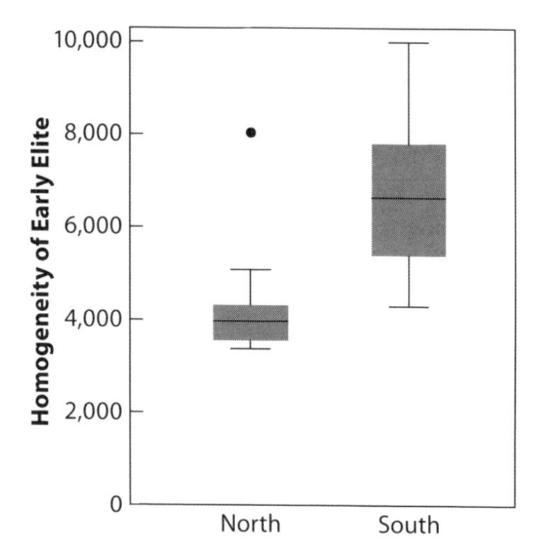

Figure 4.15. Occupational Homogeneity in 1860 in the North and the South.
In each box, the central bar is the median state, the box contains the interquartile range, and the T-sticks contain the entire population when no state is more than two standard deviations above or below average. When a state is more than two standard deviations above or below the average, it is depicted with a black circle that lies outside the T-sticks.

the effects of occupational homogeneity varied over time, but were sizeable over the period 1866–1978 and statistically significant in all but one period.

In sum, the main results showing a causal relationship between occupational homogeneity and later levels of political competition withstand the inclusion of the percentage of the population in slavery in 1860 as an additional control. This is not surprising, for two reasons. The prior results all included controls for temperature and precipitation that in part captured the effects of slavery and the Civil War. Occupational homogeneity was not exclusively a southern phenomenon.

Conclusion

Chapters 3 and 4 provide a coherent framework for understanding a large number of disparate facts from the historical and political literatures. Some state initial conditions were thought to be related to state political competition, but which initial conditions and how they were related to competition was not clear. Tensions between types of elites were recognized, but its general

TABLE 4.8.
TWO-STAGE LEAST SQUARES ESTIMATES OF OCCUPATIONAL HOMOGENEITY OF EARLY ELITES ON STATE RANNEY INDEX WITH CONTROLS FOR SLAVERY

	Ranney	*Panel A: California excluded*						
Dependent variable	*1866–2000*	*1866–1878*	*1880–1898*	*1900–1918*	*1920–1938*	*1940–1958*	*1960–1978*	*1980–2000*
Log of occupational homogeneity	−77.91***	−110.8***	−99.06***	−74.26*	−64.28	−83.87**	−79.14***	−45.55
	[−411.3, −48.0]	[−868.3, −50.0]	[−522.8, −62.4]	[−310.1, −1.0]	[−283.8, 6.7]	[−407.0, −30.3]	[−446.6, −45.4]	[−367.0, 2.9]
Log of precipitation	42.96*	73.23*	56.45***	63.19*	59.28*	55.94*	33.90	−21.83
	(21.69)	(39.05)	(18.87)	(35.08)	(30.05)	(27.73)	(20.50)	(20.14)
Log of temperature	20.39	182.1**	65.70	14.48	−0.266	43.93	−53.11	−29.19
	(44.50)	(79.64)	(57.06)	(48.02)	(51.92)	(56.27)	(44.79)	(49.97)
Log of distance to ocean	9.478**	0.916	10.09*	10.37	10.21*	7.507	14.03***	8.586
	(4.381)	(10.39)	(5.474)	(6.364)	(5.526)	(6.235)	(4.849)	(5.676)
Log of slave share	−6.581*	−7.971	−5.759	−6.794	−6.792	−10.77**	−6.820	−2.601
	(3.605)	(8.320)	(3.747)	(5.650)	(5.206)	(4.845)	(4.404)	(5.728)

	Ranney	Panel B: California included						
	1866–2000	1866–1878	1880–1898	1900–1918	1920–1938	1940–1958	1960–1978	1980–2000
Log of occupational homogeneity	−69.32**	−120.2***	−88.82**	−72.75	−53.54	−61.16	−58.73*	−49.31
	[−4368.5, −36.2]	[−14348.3 −51.9]	[−5676.4, −49.1]	[−2930.7 36.4]	[−1364.4, 150.3]	[−1741.6, 91.0]	[−58.7, −12.4]	[−5442.9, 5.1]
Log of precipitation	33.67**	83.38**	45.37**	61.56**	47.65**	31.37	11.83	−17.76
	(13.77)	(30.83)	(17.44)	(24.22)	(19.37)	(20.73)	(15.65)	(16.97)
Log of temperature	28.65	173.1**	75.56*	15.94	10.08	65.79	−33.46	−32.81
	(32.30)	(68.98)	(42.48)	(42.42)	(41.53)	(42.25)	(27.10)	(42.13)
Log of distance to ocean	8.525**	1.958	8.955*	10.20*	9.020*	4.984	11.77***	9.003*
	(3.532)	(9.880)	(4.620)	(5.272)	(4.960)	(5.913)	(3.901)	(5.075)
Log of slave share	−7.653**	−6.798	−7.038**	−6.983	−8.134	−13.60***	−9.369**	−2.132
	(3.085)	(8.184)	(3.108)	(5.677)	(4.814)	(4.112)	(3.429)	(5.576)

Notes: Top panel drops California and contains 27 observations; bottom panel keeps California and there are 28 observations. The first stage F-statistics in the top and bottom panels are 9.0 and 5.1, respectively. Since they are lower than 10, the 90 percent conditional likelihood confidence interval is reported (see Moreira 2003) in brackets under the point estimates for the log of HHI. All other cells contain point estimates and heteroskedasticity robust standard errors (in parentheses). The constant is estimated but not reported. The notations *, **, and *** denote significance at the 10 percent, 5 percent, and 1 percent levels. Log of slavery is the percentage of state population in 1860 that was enslaved, and the source is Mitchener and McLean (2000).

importance was not widely understood. The wealth of the elite was thought to play a role. Persistence in political competition was not well documented. Relative levels of political competition were partially obscured by shifts in absolute levels of political competition. The dominant narrative focused on slavery and the Civil War as the reason for low levels of political competition in the South and the dominance of the Democratic Party as the mechanism through which competition remained low. The North received relatively little attention.

Drawing on large amounts of quantitative and qualitative historical evidence covering the forty-eight states in the continental United States, the preceding chapters showed which initial conditions were related to state political competition and why. The characteristics of the elite in 1860 were documented, and the occupational homogeneity of the state elite was shown to be related to state precipitation and state distance to internal water. Moreover, occupational homogeneity of the state elite was shown to be persistently and causally negatively related to political competition in the state legislature. Wealth, in contrast, exhibited a varying relationship to political competition. The framework also provides an explanation for persistence—economic and political forces acted to create persistence in the party affiliation of legislators from individual geographic units. Finally, accounting for the effects of slavery on state political competition suggests that both slavery and occupational homogeneity of the elite played roles in shaping political competition. The point is not to dismiss the importance of slavery or the experience of the southern states. It is to show that other factors, notably the occupational homogeneity of the elite, also shaped political competition in state legislatures and political institutions more broadly.

Appendix

TABLE 4.1A.
MEASURES OF THE ELITE BY STATE

State	*Occupational homogeneity of elite, 1860*	*Wealth share of elite, 1860*
Alabama	7,080	0.244
Arkansas	10,000	0.341
California	3,910	0.289
Connecticut	4,300	0.297
Georgia	6,420	0.281
Illinois	3,380	0.321
Indiana	5,080	0.161
Iowa	4,450	0.195
Kentucky	5,010	0.256
Louisiana	5,400	0.455
Maine	4,140	0.191
Maryland	3,600	0.285
Massachusetts	4,040	0.300
Michigan	3,400	0.159
Mississippi	7,810	0.183
Missouri	3,660	0.239
New Hampshire	4,320	0.187
New Jersey	4,140	0.256
New York	3,540	0.342

TABLE 4.1A. *(continued)*

State	*Occupational homogeneity of elite, 1860*	*Wealth share of elite, 1860*
North Carolina	5,610	0.275
Ohio	3,420	0.240
Pennsylvania	3,550	0.273
South Carolina	10,000	0.222
Tennessee	4,290	0.259
Texas	6,850	0.259
Vermont	8,020	0.131
Virginia	4,750	0.331
Wisconsin	3,380	0.150

Table 4.2A.
Time Varying Effects of Occupational Homogeneity

	Cross-section	
Dependent variable	*Ranney (1900–1918)*	*Ranney (1900–1918)*
Log of occupational homogeneity	−60.52*** (10.70)	−48.37*** (13.57)
Log of slave share	X	−3.55 (2.51)
Observations	28	28

	Panel analysis (base period is 1900–1918)	
Dependent variable	*Ranney*	*Ranney*
Log of occupational homogeneity × 1866–1878	30.75** (13.21)	24.52 (20.34)
Log of occ. homogeneity × 1880–1898	3.90 (7.28)	0.86 (10.01)
Log of occ. homogeneity × 1920–1938	3.77 (4.98)	7.71 (5.43)
Log of occ. homogeneity × 1940–1958	−14.64 (9.66)	1.29 (9.16)
Log of occ. homogeneity × 1960–1978	−14.17 (13.52)	12.57 (15.61)
Log of occ. homogeneity × 1980–2000	20.75 (14.82)	34.88** (21.06)

TABLE 4.2A. *(continued)*

Dependent variable	*Panel analysis (base period is 1900–1918)*	
	Ranney	*Ranney*
State FE	Y	Y
Period FE	Y	Y
Period FE × log of slave share	N	Y
Observations	196	196
Number of states	28	28

Notes: The dependent variable in the panel analysis is the average value of the state Ranney at approximately 20-year intervals: 1866–1878, 1880–1898, 1900–1918 . . . 1980–2000. Standard errors are bootstrapped (500 repetitions) and clustered at the state level. The notations ** and *** denote significance at the 5 percent and 1 percent levels. The constant is estimated but not reported.

CHAPTER 5

State Courts

THIS CHAPTER AND THE NEXT CHAPTER examine how a state's colonial legal system and levels of political competition in the state legislature shaped the independence of judges on the state high court. Figure 5.1 illustrates the basic relationships.

The independence of judges in state high courts influences their behavior on the bench and thus economic and social outcomes. Using a sample of almost all state supreme court cases from 1995 to 1998, Shepherd (2009) found that judges facing Republican electorates in partisan reelections were "more likely to vote for businesses over individuals, for employers in labor disputes, for doctors and hospitals in medical malpractice cases, for businesses in products liability cases, for the original defendant in torts cases, and against criminals in criminal appeals."[1] Judges facing Democratic electorates had the opposite voting patterns. A variety of other studies examining different time periods and data sets have reached similar conclusions.[2]

Recent scandals have highlighted the long-standing problems associated with electing high court judges. Getting elected is expensive and requires the support of wealthy donors. These donors, not surprisingly, expect something in return. A detailed study of Ohio by the *New York Times* in 2006 suggests that donors were getting something. In split-decision cases involving contributors over the period 1994–2006, six of the ten state supreme court judges sided with contributors at least 70 percent of the time. Ohio Supreme Court justice Paul Pfeifer commented, "I never felt so much like a hooker down by the bus station in any race I've ever been in as I did in a judicial race. Everyone interested in contributing has very specific interests. They mean to be buying a vote."[3] The problem is not new—getting elected always required having well-positioned supporters, and these supporters always expected something in return for their support.

A theory based on Landes and Posner (1975) and more formally modeled by Maskin and Tirole (2004) and Hanssen (2004b) links levels of political

[1] Shepherd (2009), 188.

[2] See also Hall (1987, 1992),Tabarrok and Helland (1999), Hanssen (1999), Bohn and Inman (1996), and Huber and Gordon (2004).

[3] Liptak and Roberts (2006).

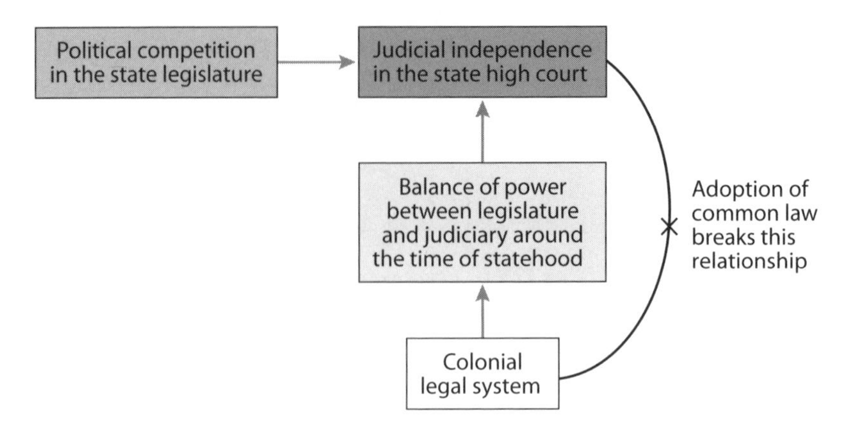

Figure 5.1. Relationships among Legal Initial Conditions, Political Competition, and State Courts.

competition in the state legislature to judicial independence. The intuition of the model is straightforward. When political competition is weak, the majority party in a legislature prefers judicial elections. Elections allow the majority party to put political pressure on judges to behave in ways preferred by the majority party. When political competition is strong, the majority party may become the minority party in a future election. In this case, the majority prefers to appoint judges, because they are likely to make favorable rulings in the future if a new majority party emerges.

Independence of state high court judges has been highly persistent. The model provides an explanation for this persistence. Changes will only be made to judicial retention procedures if there are substantial changes in the levels of political competition in the state legislature. In the absence of changes in the levels of political competition, current levels of judicial independence will persist.

Drawing on chapter 2, the model is extended to allow civil-law and common-law legislatures to differ in the benefits they receive from more independent judges. If the benefits to civil-law legislatures of judicial independence are lower, the threshold level of political competition at which they will increase judicial independence will be higher. It is worth noting that for twelve of the thirteen civil-law states shown in figure 5.2, the adoption of common law arguably broke the direct link between the colonial legal system and state courts. The remaining state—Louisiana—poses a problem, because it has not adopted common law. Therefore, our analysis of state courts excludes Louisiana.

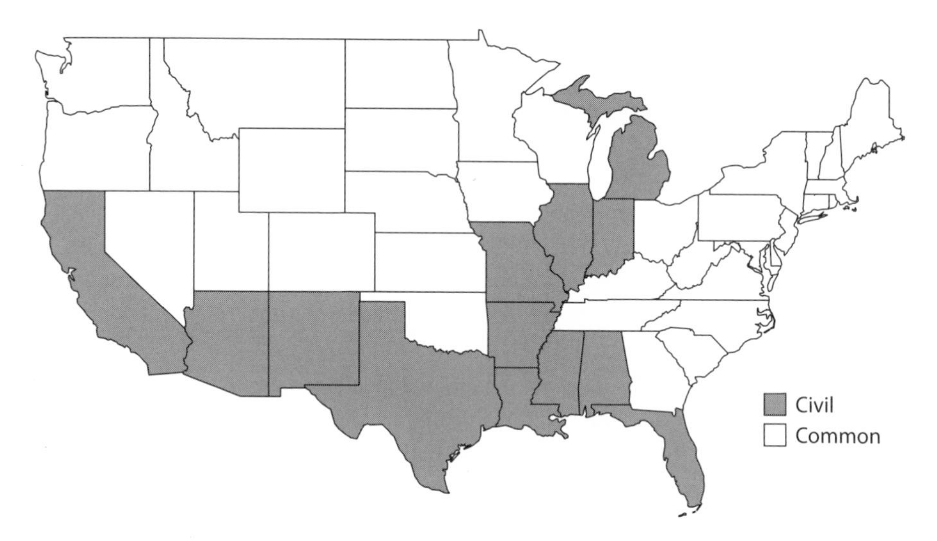

Figure 5.2. Colonial Legal System.

Empirically, civil-law and common-law states have differed in ways predicted by the theory. Civil-law states had lower levels of judicial independence than common-law states, even when controls are included for the timing of the entry of the state into the union and for slavery and when alternative measures of civil law are used. Moreover, civil-law states increased judicial independence at higher levels of political competition than common-law states.

Civil-law states also differed from common-law states in their adoption of intermediate appellate courts. Intermediate appellate courts allow better monitoring of lower-court judges, because many more cases can be reviewed. Under civil law, judges are generally monitored much more closely than under common law. Consistent with this pattern, civil-law states adopted intermediate appellate courts at much lower population levels than common-law states.

Judicial Selection and Retention

One of the most important structural features of a state court system is how the supreme court justices are selected and retained. The focus here is on supreme court justices for two reasons. State supreme court justices ultimately hear the most difficult and important state-court cases. Further, judges on the next lower tier of state courts are often selected and retained in the same way as state high court judges.

The relative importance of selection and retention for judicial behavior is the source of some debate.[4] In practice, it may be difficult to separate the effects of the two, because selection and retention mechanisms tend to be highly correlated. For example, judges who are selected in elections will almost always be retained through elections. We will follow most of the literature and focus on retention procedures. Retention procedures shape the extent to which a judge is free to make decisions without regard to outside pressures. This freedom is often referred to as judicial independence.[5]

Elections tend to offer the least judicial independence, because judges need to be reelected to retain their positions. Under partisan elections, party officials determine whether judges are put on the ballot and control the monetary and nonmonetary resources necessary for reelection. To be reelected, judges have to please both party officials and voters. The need to please these constituents is likely to affect decisions in complicated or controversial cases. Nonpartisan elections are considered to offer somewhat more independence than partisan elections. In general, nonpartisan elections tend to require less cultivation of party officials. In some cases they also require less fund-raising.

Numerous scholars and officials have publicly opposed the election of judges. In a 1906 address to the American Bar Association, the renowned legal scholar Roscoe Pound argued that "putting courts into politics and compelling judges to become politicians in many jurisdictions . . . [has] almost destroyed the traditional respect for the bench."[6] The American Bar Association is on record as opposing both partisan and nonpartisan judicial elections. "The American Bar Association urges state, territorial, and local bar associations in jurisdictions where judges are elected in partisan or non-partisan elections to work for the adoption of merit selection and retention."[7]

[4] A number of authors have argued that the judges selected are similar, but they behave differently once on the bench, because of the retention process. Canon (1972), Glick and Emmert (1987), and Besley and Payne (2003) address this issue. Hall (1984) provides evidence that northern and southern judges have similar educational and family backgrounds. Recent empirical work by Choi, Gulati, and Posner (2008) suggests that both may be important. Judges subject to partisan elections were more likely to have gone to in-state and lower-ranked law schools than appointed judges. They argue that elections encourage state judges to behave more like politicians who focus on providing service to their constituents and that appointment to the judiciary encourages judges to behave more like professionals who focus on building a legacy as creators of law.

[5] See Kornhauser (2002) for a discussion of the concept and its usefulness.

[6] Pound (1906).

[7] American Bar Association resolution in 1994, quoted in Cardman (2007).

Retention elections and appointment-based systems are generally thought to offer considerably more judicial independence. In a retention election, judges run unopposed. Voters vote yes or no on questions that effectively ask: "Should Judge X be retained?" Judges are almost always retained.[8] The highly politicized and publicized failure of three California Supreme Court justices to be retained in 1986 made it clear that retention elections carry some risk. Appointment-based systems also offer some risk that a judge will not be reappointed, particularly if there has been a change in the party in power or if the court has made one or more very unpopular decisions. In both cases, however, the risk tends to be low relative to the risk associated with elections.[9]

Other features of the court system, such as term length, also affect judicial independence. Judges with longer term lengths are relatively more insulated than judges with shorter term lengths. They can make unpopular decisions early in their career and have time to recover. Depending on their age when they move onto the supreme court, they may choose not to pursue a second term.

Judicial retention procedures vary across states and have varied at the state level over time. The standard narrative divides historical trends in the election and retention of judges into four periods.[10] During the earliest period (1790–1847), all judges were appointed by the legislature or governor or the two acting jointly. This retention process reflected a number of factors, including the primacy of early legislatures, a lack of distinction between lawmaking and judging, and a distrust of colonial judges, many of whom had been loyal to the Crown.

During the second period (1847–1910), states adopted partisan elections in response to the perceived need for state courts to be independent of state legislatures. Based on case studies of four states, Kermit Hall (1987) found that "From the outset, popular election of judges was a lawyers' reform, one in which politically active lawyers in California and elsewhere dominated the constitutional

[8] See Dubois (1980), Hall (1987), Hall (1995), Brace and Hall (1997), and Bonneau (2005).

[9] See Dubois (1980) for the period 1948–1974 in nonsouthern states. In table 13, p. 109, he shows that sizeable numbers of incumbents were defeated in some states with elections. Ohio (10), Colorado (6), Michigan (5), New Mexico (5), and Indiana (4) all had sizeable numbers of defeats. Many of these judges were appointed to finish an unexpired term and then were defeated in their first election. See Bonneau (2005) for the period 1990–2000. He finds 39 percent of incumbents were defeated in contested partisan elections and 10 percent were defeated in contested nonpartisan elections. Hall (2001) finds similar results for the period 1980–1995. Hall (1987) finds that over the period 1850–1920 18 percent of California Supreme Court justices were not renominated by their party or were not reelected. During the period from 1879–1900, the rate was 29 percent. Hall (2006) finds that over this same period 20 percent of Ohio Supreme Court justices were defeated at the polls.

[10] See Hanssen (2004a) and Epstein, Knight, and Shvetsova (2002).

conventions that adopted partisan elections. These lawyers conceived of election as a means of upgrading the judiciary—and the entire legal profession—by providing it with a base of popular support."[11] Twenty of the twenty-nine existing states and all seventeen of the new states adopted partisan elections. Unfortunately, partisan elections forced judges to participate in the same processes as other political actors, leading to many of the same problems.

In the third period (1910–1958), states tried to reduce the politicization of the judiciary by adopting nonpartisan elections. Seventeen of the forty-six existing states and one of the two new states adopted nonpartisan elections. Although nonpartisan elections were perceived to be an improvement, many citizens felt that judges were still inadequately insulated from the political process. This was the case, for example, in Missouri in the 1930s, where the abuses of the state court system by "Boss Tom Pendergast" in Kansas City and ward bosses in St. Louis led to intense public backlash.[12]

In the fourth period (1958–present), states moved to a system of unopposed retention elections, coupled with merit selection of judges. In 1940 Missouri introduced a merit system for many judges, including state supreme court justices.[13] Under the "Missouri plan," a nonpartisan and expert commission selected candidates for a judicial vacancy based on merit.[14] Once the selected candidate finished his or her term, he or she stood for uncontested retention election. The next adoption of the Missouri plan did not occur until 1958, when Kansas made the switch. By 1990, fourteen of the forty-eight continental states had both merit selection and retention elections, and three additional states had other methods of selection coupled with retention elections.[15]

It is worth noting that the American Bar Association was not entirely a disinterested party to adoption of the merit system. Lawyers have benefited from the merit systems in two ways. They typically have greater input into who becomes a judge than under other systems. Further, because independent judges are less predictable, it is more worthwhile to file cases than it was when judges were more predictable. In states that adopted merit plans, Hanssen (2002) found that the number of cases filed in state supreme courts increased by 18–32 percent following adoption.

[11] Hall (1987), 65.

[12] See Karlen (1970).

[13] California adopted retention elections in 1934, but it did not have merit selection. Thus, Missouri is considered to have introduced the first true merit plan.

[14] The governor or the judicial commission selects the final candidate.

[15] *The Book of the States* (1991). For case studies of change and failure to change, see Becker and Reddick (2003).

As was discussed in the introduction, different methods of retention appear to have affected judicial behavior. Whether this holds before the 1970s is an open question, because so little work has been done on the earlier period. Shugerman (2009) examined judicial response to the Johnstown flood of 1889, which killed 2,000 people. The flood occurred when heavy rain led to the collapse of a poorly maintained dam. Under the prevailing doctrine, the owners of the dam were not strictly liable for the disaster. Elected judges tended to adopt strict liability for "unnatural activities" quite rapidly after the flood, while appointed judges were much slower. Shugerman argues that the response of elected judges was in part due to selection effects. The election process tended to filter out elite professionals and simultaneously attract lawyer-politicians. The latter tended to be much more responsive to public opinion. He also finds evidence suggesting that term length affected judges' response to the disaster. Elected judges who had shorter terms needed to please a variety of constituents, including special interests and party officials. Elected judges with longer terms tended to be the most responsive to public opinion. Although they were not required to take action, their backgrounds as lawyer-politicians led them to adopt strict liability.

A Model of State Legislatures and State Courts

A line of work that stretches back to Landes and Posner (1975) and has been developed more fully by Epstein, Knight, and Shvetsova (2002), Maskin and Tirole (2004), and Hanssen (2004a) argues that changes in judicial retention systems are related to changes in levels of political competition. Hanssen (2004b) provides a model of how political competition influences judicial retention. This section incorporates civil law into Hanssen's model. Since the general intuition has already been discussed in the introduction, readers can skip this section without missing the general argument.

Consider a world with three periods denoted 0, 1, and 2. In periods 0 and 1, Party A is the majority party in the legislature. At the end of period 1 there are legislative elections. Party A wins the election with probability $x > 0.5$, and Party B wins with probability $1 - x$. The party that wins at the end of period 1 is the majority party in period 2.

In period 0, Party A chooses a system of judicial elections or appointments. Appointed judges are guaranteed a job for periods 1 and 2. Elected judges have a job in period 1 and, if reelected, in period 2. If the elected judge loses after period 1, the majority party in period 2 picks a new judge, who then makes a ruling.

In period 1 and in period 2, a judge rules either "a" or "b." Party A receives a payoff equal to 1 when a judge rules "a" and a payoff equal to 0 when a judge rules "b." Party B has a completely different agenda and receives a payoff of 1 when a judge rules "b" and a payoff of 0 when a judge rules "a."

At the beginning of period 1, Party A picks a judge, where $\delta > 0.5$ is the probability that the chosen judge prefers ruling "a"; δ can be thought of as representing a party's ability to screen for judges who will behave in the way that the party prefers. The judge then makes a ruling that is observed at the end of the period. Suppose that the judge is elected. If the incumbent party, Party A, wins the election and the judge rules "a" in period $t = 1$, that judge remains in office. If Party A wins and the judge rules "b" in period 1, the judge is removed, and Party A picks a new judge at the start of period $t = 2$. If Party B wins, the judge keeps her job in period 2 only if she ruled "b" in period 1. Since x is the probability that Party A remains in office in period 2, our measure of political competition, denoted pc, is $1 - x$.

There are two kinds of judges, and their preferences are common knowledge. Let P denote the payoff to a judge who makes her preferred ruling in period 1, and let O denote her payoff from holding onto her office during periods 1 and 2. If judges are elected and O > P, then the judge makes ruling "a" in period 1 and makes her most preferred ruling in period 2. In this case, the legislature's threat of being able to remove her from office gets the judge to make ruling "a" in period 1 even when "b" is her preferred ruling. If O < P, the judge makes her preferred ruling in period 1 even if this means she loses her job. In period 2, a judge makes her preferred ruling. Setting up a simple game tree, it is straightforward to show Party A's payoffs associated with elections when O < P and O > P are

$$\text{V, elections (if O} < \text{P)} = \delta(2 + x - \delta)$$
$$\text{V, elections (if O} > \text{P)} = 1 + x\delta + (1 - x)(1 - \delta).$$

Now suppose that civil-law legislatures experience disutility from an independent judiciary, denoted c, where $0 < c < 1$. Then it is straightforward to show that Party A's payoffs from a system of appointments when it operates in a common-law and civil-law state are

$$\text{V, appointed (common)} = 2\delta$$
$$\text{V, appointed (civil)} = 2(\delta - c).$$

Using these payoffs, critical levels of political competition can be computed, $pc^* = 1 - x^*$, for common-law and civil-law legislatures.

Suppose O < P, which means that elected judges make their most preferred ruling, even if this costs them their job. Thus, for common-law states, the critical level of political competition is $pc^*(\text{common}) = 1 - \delta$. For civil-law states, it is $pc^*(\text{civil}) = 1 + 2c - \delta$. Civil-law legislatures require a higher level of political competition before they switch to an appointment system. If screening is sufficiently precise (δ is close to unity), common-law states would almost always prefer appointments, and political competition will be less important as a force for removing elections.

Suppose that O > P, which means elected judges make the ruling "a" in period 1 so that they can stay in office. Then, $pc^*(\text{common}) = (1 - \delta) / (2\delta - 1)$ and $pc^*(\text{civil}) = (1 - \delta + 2c) / (2\delta - 1)$. Again, civil-law states require higher political competition to make changes.

If screening is assumed to be constant across states, the model implies that for any political history where the maximum level of political competition for all states is below $pc^*(\text{common})$, all states will have elections. Similarly, for any political history where the maximum level of political competition for all states is above $pc^*(\text{civil})$, all states will have elections. In the interval $pc^*(\text{common})$ to $pc^*(\text{civil})$, common-law states will appoint their judges, and civil-law states will elect their judges.

It is worth noting two other things about the model. Although the discussion focuses on how increases in political competition will cause states to switch from elections to appointment, the reverse can also happen. That is, declines in political competition can lead states to switch from appointment to elections. The model also implies that if political competition in a state varies around pc^*, a state might switch frequently, possibly in every period. In practice, changes in judicial retention procedures for high court judges involve changing the state constitution, which is time consuming and costly. So states are unlikely to make frequent changes to judicial retention procedures. Indeed, most states have made between zero and two changes to judicial retention procedures over the course of their history.

Political Competition, Civil Law, and Judicial Retention

Table 5.1 presents summary statistics for judicial retention systems and political competition. The average value of partisan elections was 0.46, the average value of judicial independence was 2.9, and the average term length was 9.6 years, and the average value of the Ranney index was 52. The coding of judicial independence is from to Epstein, Knight, and Shvetsova (2002). Their index runs from 1, which is partisan elections and low judicial independence,

TABLE 5.1.
SUMMARY STATISTICS FOR JUDICIAL RETENTION AND POLITICAL COMPETITION

	Partisan	*Judicial independence*	*Term lengths*	*Intermediate appellate courts*	*Ranney index*
Average	0.460	2.93	9.58	0.263	51.99
Standard deviation	0.487	2.57	5.35	0.430	28.14
Minimum	0	1	1	0	0
Maximum	1	9	25	1	96.84
Number of states that change	33	40	24	35	46

Sources: See the appendix to this chapter.

Notes: Louisiana is excluded because it retained the civil law after entering the union, and Nebraska is dropped because it has a unicameral legislature for most of 1866–2000. This leaves 46 states in the sample, and, consistent with subsequent tables, decadal averages are used. There are 608 observations of partisan, judicial independence, and term lengths; and, there are 616 observations of the Ranney index. Results are similar if we use annual data. A term length of 25 years denotes life term tenure.

to 9, which is life tenure and high judicial independence.[16] It tries to capture the opportunity cost of a judge acting on his beliefs in a potentially controversial case. Judges with higher opportunity costs have lower scores and less judicial independence.

Table 5.2 shows the correlations among judicial retention systems, political competition, and colonial legal system. Use of partisan elections was negatively correlated with the Ranney index and positively correlated with civil law. Judicial independence was negatively correlated with civil law and relatively uncor-

[16] Higher numbers correspond to fewer parties being involved in the reappointment process. The coding is partisan elections (1), nonpartisan elections (2), retention elections (3), governor, legislature, and commission reappointment (4), governor and legislature reappointment (5), reappointment by both houses (6), governor and commission reappointment (7), commission reappointment (8), life tenure (9).

TABLE 5.2.
CORRELATIONS FOR JUDICIAL RETENTION AND POLITICAL COMPETITION

	Partisan	*Judicial independence*	*Term length*	*Ranney*	*Intermediate appellate courts*
Partisan	1.000				
Judicial independence	−0.687	1.000			
Term length	−0.270	0.689	1.000		
Ranney	−0.272	0.066	0.145	1.000	
Intermediate appellate courts	0.016	−0.198	−0.115	0.181	1.00
Civil	0.287	−0.286	−0.145	−0.144	0.311

Sources: See the appendix to this chapter.

Notes: Louisiana is excluded because it retained the civil law after entering the union, and Nebraska is dropped because it has a unicameral legislature for most of 1866–2000. This leaves 46 states in the sample, and, consistent with subsequent tables, decadal averages are used.

related with the Ranney index. Term length was weakly positively correlated with the Ranney index and weakly negatively correlated with term length.

Figure 5.3 shows that levels of judicial independence were highly persistent over time. States with higher levels of judicial independence in 1900–1918 had higher levels of judicial independence from 1866 to 2000. This is striking and suggests that levels of judicial independence were set very early on.

The model predicts that the threshold levels of political competition above which states will change their retention methods will differ in civil-law and common-law states. This implies that civil-law states will, controlling for the level of political competition, be more likely to use partisan elections and have lower judicial independence than common-law states.

Figure 5.4 illustrates how the evolution of political competition was related to the evolution of partisan election systems. Recall that partisan elections were widely considered to be good for the quality of courts through at least 1900. The model predicts that partisan elections will be removed when political

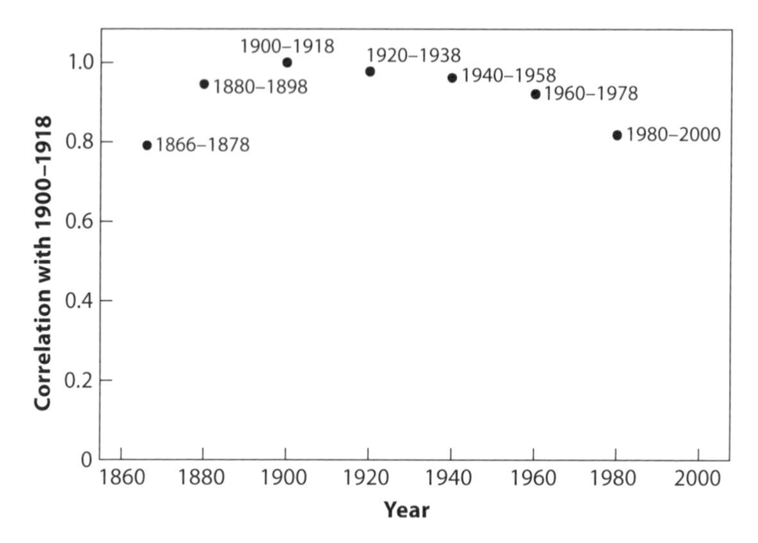

Figure 5.3. Persistence of Judicial Independence in State High Courts' Judicial Independence, 1866–2000.

The judicial independence index runs from 1 (partisan elections and least independent) to 9 (lifetime tenure and most independent). This index was constructed by Epstein, Knight, and Shvetsova (2002) and is discussed in detail in chapter 5. Louisiana is excluded because it retained civil law after entering the union and Nebraska is dropped because it has a unicameral legislature for most of 1866–2000. Eleven additional states are dropped for lack of data. This leaves 35 states in the sample. The results are similar if we include these 11 states and conduct the analysis for 1910–2000.

competition is sufficiently high. The trends starting in the mid-1920s are consistent with this prediction.

Figure 5.5 confirms that common-law and civil-law states differed systematically in their use of partisan elections. Civil-law states adopted partisan elections more rapidly than common-law states during the nineteenth and early twentieth centuries. Civil-law states that had partisan elections moved away from them more slowly during the twentieth century than did common-law states.

Why would civil-law states have been so enamored with partisan elections and elections more generally? Following the French Revolution, the French briefly considered using partisan elections to retain judges. Ultimately, judges were made part of a bureaucracy.[17] Both historically and today, French judges,

[17] See Bell (1988), Volcansek et al. (1996), and Merryman and Perez-Perdomo (2007).

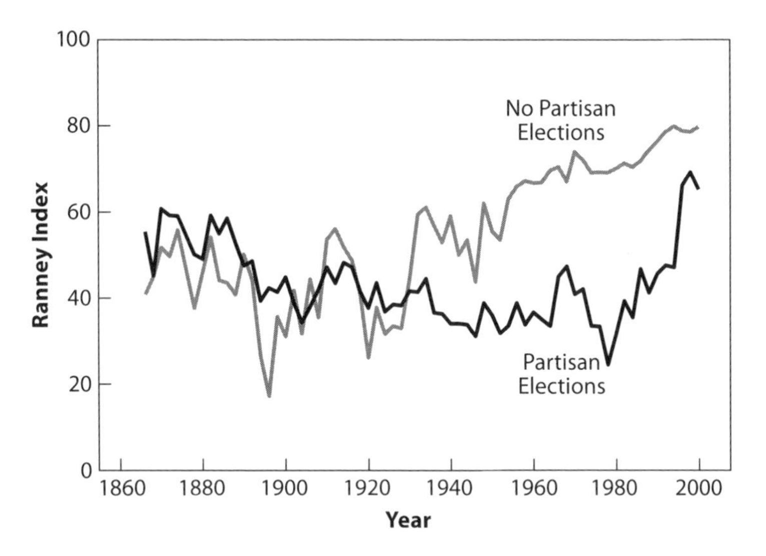

Figure 5.4. Political Competition and Partisan Elections, 1866–2000.

Louisiana is excluded because it retained civil law after entering the union, and Nebraska is dropped because it has a unicameral legislature for most of 1866–2000. This leaves 46 states in the sample. *Sources*: See the appendix to this chapter.

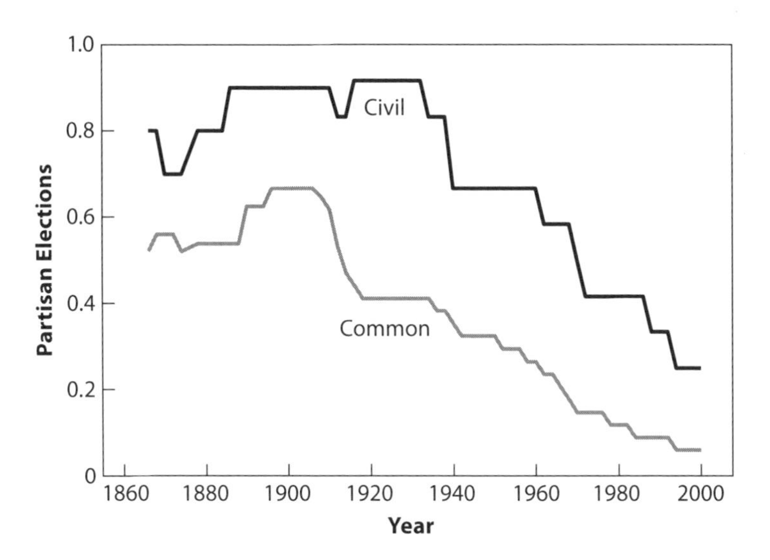

Figure 5.5. Use of Partisan Elections for High Court Judges by Common-Law and Civil-Law States, 1866–2000.

The vertical axis is the share of states that use partisan elections. Louisiana is excluded because it retained civil law after entering the union, and Nebraska is dropped because it has a unicameral legislature for most of 1866–2000. This leaves 46 states in the sample. *Sources*: See the appendix to this chapter.

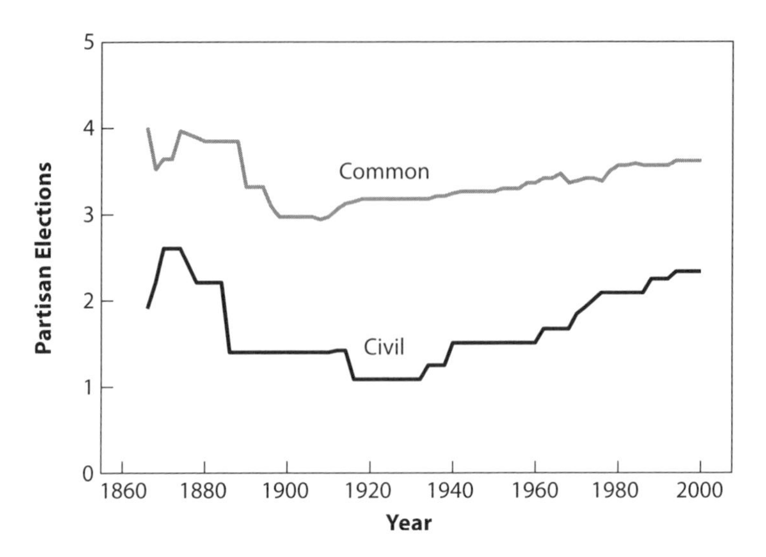

Figure 5.6. Epstein et al. Measure of Judicial Independence, 1866–2000.

The Epstein, Knight, and Shvetsova (2002) index runs on a scale of 1 (least independent) to 9 (most independent). The vertical axis is the average score on the Epstein et al. index for civil-law and common-law states. Louisiana is excluded because it retained civil law after entering the union, and Nebraska is dropped because it has a unicameral legislature for most of 1866–2000. This leaves 46 states in the sample.
Sources: See appendix to this chapter.

particularly younger judges, are supervised and can be punished for decisions that systematically differ from norms regarding appropriate outcomes. Adherence to the norms is an important factor in promotion within the bureaucracy. Outside options are limited, so the primary path is to rise within the judicial hierarchy. This system is in many ways more restrictive than the American system under partisan elections, because judges are constantly being evaluated and can be sanctioned for poor decision making at any time, not just at the time of reelection.

Figure 5.6 documents differences in judicial independence in common-law and civil-law states. Over the period 1866–2000, civil-law states had systematically lower levels of judicial independence than common-law states.

The evidence presented in figures 5.4–5.6 is suggestive but hardly conclusive. Table 5.3 examines the relationships among political competition, colonial legal origins, and judicial retention in a regression framework. Because retention

TABLE 5.3.
CIVIL LAW, POLITICAL COMPETITION, AND PARTISAN ELECTIONS

Dependent variable	*Partisan elections 1866–2000*	*Partisan elections 1866–2000*	*Partisan elections 1866–2000*	*Partisan elections 1866–2000*
Civil	0.280***	0.253**	0.108	
	(0.0880)	(0.108)	(0.0794)	
Log of duration civil law				0.0236
				(0.0184)
Ranney	–0.0040***	–0.0001	–0.0024*	–0.0023*
	(0.0013)	(0.0014)	(0.0012)	(0.0012)
Pre-1847		–0.151	0.134	0.134
		(0.0937)	(0.0991)	(0.0983)
Log of slavery		0.0739*	0.0241	0.0243
		(0.0402)	(0.0337)	(0.0335)
State FE	N	N	N	N
Decadal FE	N	Y	Y	Y
Sample	Full	Full	Switchers	Switchers
States	46	46	33	33
Observations	607	607	430	430
R^2	0.136	0.294	0.418	0.417

Notes: Standard errors are bootstrapped (500 repetitions) and clustered at the state level. The notations *, **, and *** denote significance at the 10 percent, 5 percent, and 1 percent levels. The constant is estimated but not reported. Nebraska is always dropped from the sample because it has a unicameral legislature. Louisiana is always dropped because it retained the civil law after joining the union. Partisan equals 1 when partisan elections are in force and 0 otherwise. Because retention procedures change infrequently, partisan elections are averaged over 10-year periods. Switcher states changed at least once during 1866–2000. Log of duration of civil law is the average of the upper and lower bounds for duration of civil law reported in table 2.4.

systems change infrequently, measures of Ranney and retention systems are decadal averages.

The first column suggests that the basic intuition of the model is correct. States with higher levels of competition in the state legislature were statistically significantly less likely to use partisan elections, and civil-law states were statistically significantly more likely to use partisan elections. The effects of political competition were relatively small. A twenty-point increase in the Ranney index decreased the predicted use of partisan elections by 0.08.[18] The effect of civil law was substantially larger. Civil-law states' predicted use of partisan elections was 0.28 higher than common-law states'.[19]

One concern is that the civil-law variable may be picking up things other than the historical influence of civil law on the balance of power between state legislatures and state courts. For example, since many civil-law states are in the South, civil law could be capturing something related to slavery. Or it could be capturing differences between states that entered before 1847, when all states entered with appointment, and states that entered after 1847, when states entered with partisan or nonpartisan elections.

An important decision with respect to specification is whether to include decadal fixed effects. The model focuses exclusively on the state legislature, but other factors may affect whether a change in retention is likely to occur. For example, outside actors may lobby for change, the governor may or may not support change, and the electorate typically has to ratify the change. Further, the economy and changing views on the appropriate system for judicial retention may also affect change. The inclusion of decadal fixed effects will capture national trends likely to affect changes in retention systems. At the same time, their inclusion poses a much more restrictive test for the importance of the Ranney index, because it will also capture national time trends in the Ranney index.

The remaining columns in table 5.3 include controls. Adding controls has little effect on the civil variable in the second column. The coefficient remains positive, statistically significant and large.

The last two columns restrict the sample to the thirty-three states that changed to or from partisan elections at some point during 1866–2000. In particular

[18] Twenty points is roughly two-thirds of a standard deviation in the Ranney index.

[19] Because the graphs suggest that the effect of the civil variable was relatively constant over time, its effects are estimated as a constant, rather than a time-varying effect. For more on the time-varying effects, see table 5.1A. For most outcomes in most time periods, the coefficient on the civil variable was not statistically significantly different from its value in 1900–1908.

this sample excludes Arizona, Arkansas, Connecticut, Delaware, Maine, Massachusetts, New Hampshire, New Jersey, Rhode Island, South Carolina, Vermont, Virginia, and West Virginia. The fourth column replaces civil law with a variable measuring the duration of civil law during the colonial period.[20] The effect of civil law on the use of partisan elections falls and becomes statistically insignificant when the sample is restricted to switchers. The coefficient on the Ranney index is small, but statistically significant for switchers, even with the inclusion of decadal fixed effects.

Table 5.4 moves away from partisan elections to focus on judicial independence more broadly. It shows that civil law and entry date are strongly related to judicial independence. Across all four specifications, the coefficient on civil law is negative, statistically significant, and large. Levels of judicial independence in civil-law states were 1.0–1.7 lower than in common-law states. The type of retention system that a state entered with also had an enduring effect. States that entered before 1847 had levels of judicial independence that were 1.5–2.5 higher than states that entered after.[21] These are large effects, given that the mean level of judicial independence in the sample was 2.9.

Overall, tables 5.3 and 5.4 confirmed that civil-law and common-law states used different methods to retain their state high court judges. Civil-law states were more likely to use partisan elections and had lower levels of judicial independence than common-law states. This is consistent with the predictions of the model.

We test the prediction of the model more directly in table 5.5 by comparing the threshold levels of political competition at which common-law and civil-law states switch. Levels of political competition are included in the two periods preceding the switch, since it may take time for change to occur, conditional on having reached the threshold level of political competition, pc^*. Further, the two periods following the switch are included, since the relevant political players may choose to act on change prior to actually reaching pc^*. That is, they may forecast reaching pc^* in the near future and initiate change while their political power is still strong.

[20] Duration of civil law is taken from table 2.4 and is the average of the upper and lower bounds for years of civil law.

[21] This result would appear to differ from the result in table 5.3. The difference stems in part from differences in the composition of the sample—thirty-three states changed to or from partisan elections, while forty states changed judicial independence. These seven additional states had much higher levels of judicial independence than the thirty-three that changed to or from partisan elections.

TABLE 5.4.
CIVIL LAW, POLITICAL COMPETITION, AND JUDICIAL INDEPENDENCE

Dependent variable	*Judicial independence 1866–2000*	*Judicial independence 1866–2000*	*Judicial independence 1866–2000*	*Judicial independence 1866–2000*
Civil	−1.663***	−1.422**	−1.014*	
	(0.501)	(0.685)	(0.579)	
Log of duration of civil law				−0.218*
				(0.126)
Ranney	0.0020	−0.0088	−0.0072	−0.0074
	(0.0063)	(0.0075)	(0.0081)	(0.0081)
Pre-1847		2.450***	1.459**	1.458**
		(0.748)	(0.619)	(0.612)
Log of Slavery		−0.430	−0.135	−0.138
		(0.271)	(0.260)	(0.260)
State FE	N	N	N	N
Decadal FE	N	Y	Y	Y
Sample	Full	Full	Switchers	Switchers
States	46	46	40	40
Observations	607	607	523	523
R^2	0.083	0.293	0.180	0.178

Notes: Standard errors are bootstrapped (500 repetitions) and clustered at the state level. The notations *, **, and *** denote significance at the 10 percent, 5 percent, and 1 percent levels. The constant is estimated but not reported. Nebraska is always dropped from the sample because it has a unicameral legislature. Louisiana is always dropped because it retained the civil law after joining the union. Judicial independence is an increasing ordinal scale coded by Epstein, Knight, and Shvetsova (2002) and goes from partisan elections (1), nonpartisan elections (2), retention elections (3), governor, legislature, and commission reappoint (4), governor and legislature reappoint (5), 2 houses reappoint (6), governor and commission reappoint (7), commission reappoints (8), life tenure (9). Because retention procedures change infrequently, judicial independence is averaged over 10-year periods. Switcher states changed at least once during 1866–2000. Log of duration of civil law is the average of the upper and lower bounds for duration of civil law reported in table 2.4.

TABLE 5.5.
RANNEY INDEX AT THE TIME OF CHANGES IN JUDICIAL RETENTION

	Common states	*Civil states*	*Difference in means*
Removal of partisan elections	59.4 (2.51)	67.5 (3.42)	8.19* (4.24)
Observations/states	108/22	45/9	153/31
Increasing judicial independence	57.6 (2.27)	68.8 (3.10)	11.2*** (3.84)
Observations/states	137/24	55/10	192/34
Adoption of partisan elections	38.9 (9.33)	32.4 (10.9)	−6.43 (14.3)
Observations/states	9/2	12/3	21/5
Decreasing judicial independence	68.3 (4.47)	32.4 (10.9)	−35.9*** (11.7)
Observations/states	39/7	12/3	51/10

Notes: The Ranney index and population are measured at time of reform, two and four years before and after reform. Thus, each state that makes a reform has at most five observations. A reform reverse would be either an increase of a particular indicator followed by a decrease (for example, an increase in judicial independence followed by a decrease) or an increase in terms followed by a decreased. Because a real reform has some durability, we ignore reforms that are reversed in less than eight years. Reform reversal, in fact, is a rare event in the American states. Cases include the replacement of partisan elections with nonpartisan elections in Iowa in 1912, and the reinstallment of partisan elections in 1918; and the replacement of partisan election with nonpartisan elections in Kansas in 1914 and in Montana in 1910 as and the reinstallment of partisan elections within two years. In these three cases, it is assumed that partisan elections were never eliminated. Term limit reforms are never reversed and intermediate appellate court reforms are also never reversed. Each cell reports the mean, standard error (in parenthesis), and number of observations / number of states. The column for difference in means reports a test for the null that the difference in means is zero against a two-sided alternative. The notations * and *** denote significance at the 10 percent and 1 percent levels. The test statistic allows the common- and civil-law states to have unequal variances. Several states change partisan election and judicial independence more than once. As a result, the number of states is not simply the number of observations divided by five.

Consistent with the model, civil-law states moved away from partisan elections and increased judicial independence at higher levels of political competition than common-law states. For partisan elections, the average Ranney index was 58 in common-law states and 70 in civil-law states. For judicial independence, the average Ranney index was 59 in common-law states and 71 in civil-law states. In both cases, the differences were statistically significant using a two-sided *t*-test. The states that adopted partisan elections were southern states that did so in the aftermath of the Civil War. They did so at very low levels of political competition. Six other common-law states decreased judicial independence, and they did so at relatively high levels of political competition. This creates the large differential between common-law states and civil-law states with respect to decreasing judicial independence.

Table 5.6 examines a related prediction of the model. The model implies that changes in judicial retention should be associated with increased political competition in the state legislature. The idea is that an increase in political competition pushes the legislature into a region in which the legislature prefers a more independent judiciary. By extension, this intuition should hold for both common-law and civil-law states. Our model does not, however, have direct implications regarding the relative strength of the relationship in common-law and civil-law states.

Whether the relationship between political competition and the use of partisan elections differs for the two groups of states is explored by estimating the following specification.

$$\text{Partisan}_{id} = \alpha + \beta_1 \text{Ranney}_{id} + \beta_2 \text{Ranney}_{id} {}^* \text{Civil}_i + S_i + \varepsilon_{id}.$$

Partisan_{id} is the share of state years when partisan elections were in force and the Ranney index in state *i* in decade *d,* and Ranney_{id} is the Ranney index in state *i* in decade *d*; Civil_i is an indicator variable for civil law, S_i is state fixed effects, and ε_{id} is the stochastic error term. β_1 measures the impact of the Ranney index in common-law states, and β_2 measures the differential impact of the Ranney index in civil-law states.

Table 5.6 shows that states removed partisan elections during periods of rising political competition.[22] In the first column, a twenty-point increase in the Ranney index decreased the predicted use of partisan elections by 0.12.

[22] This table and similar tables that follow do not include decadal fixed effects. Adding decadal fixed effects causes the coefficient on the Ranney index to become statistically insignificant in most specifications.

TABLE 5.6.
POLITICAL COMPETITION, CIVIL LAW, AND PARTISAN ELECTIONS WITH STATE FIXED EFFECTS

Dependent variable	*Partisan elections 1866–2000*	*Partisan elections 1866–2000*	*Partisan elections 1866–2000*	*Partisan elections 1866–2000*
Ranney	−0.0060***	−0.0084***	−0.0076***	−0.0076***
	(0.0011)	(0.0012)	(0.0015)	(0.0015)
Ranney × Civil			−0.0027	
			(0.0026)	
Ranney × log of duration of civil law				−0.0006
				(0.0006)
State FE	Y	Y	Y	Y
Decadal FE	N	N	N	N
Sample	Full	Switchers	Switchers	Switchers
States	46	46	33	33
Observations	607	430	430	430
R^2	0.560	0.390	0.393	0.393

Notes: Standard errors are bootstrapped (500 repetitions) and clustered at the state level. The notation *** denotes significance at the 1 percent level. The constant is estimated but not reported. Nebraska is always dropped from the sample because it has a unicameral legislature. Louisiana is always dropped because it retained the civil law after joining the union. Partisan equals 1 when partisan elections are in force and 0 otherwise. Because retention procedures change infrequently, partisan elections are averaged over 10-year periods. Switcher states changed at least once during 1866–2000. Log of duration of civil law is the average of the upper and lower bounds for duration of civil law reported in table 2.4.

TABLE 5.7.
POLITICAL COMPETITION, CIVIL LAW, AND JUDICIAL INDEPENDENCE WITH STATE FIXED EFFECTS

Dependent variable	*Judicial independence 1866–2000*	*Judicial independence 1866–2000*	*Judicial independence 1866–2000*	*Judicial independence 1866–2000*
Ranney	0.0079*	0.0086*	0.0027	0.0026
	(0.0045)	(0.0048)	(0.0053)	(0.0052)
Ranney × Civil			0.0215**	
			(0.0091)	
Ranney × Log of duration of civil law				0.0047**
				(0.0020)
State FE	Y	Y	Y	Y
Decadal FE	N	N	N	N
Sample	Full	Switchers	Switchers	Switchers
States	46	40	40	40
Observations	607	523	523	523
R^2	0.864	0.752	0.762	0.762

Notes: Standard errors are bootstrapped (500 repetitions) and clustered at the state level. The notations * and ** denote significance at the 10 percent and 5 percent levels. The constant is estimated but not reported. Nebraska is always dropped from the sample because it has a unicameral legislature. Louisiana is always dropped because it retained the civil law after joining the union. Judicial independence is an increasing ordinal scale coded by Epstein, Knight, and Shvetsova (2002) and goes from partisan elections (1), nonpartisan elections (2), retention elections (3), governor, legislature, and commission reappoint (4), governor and legislature reappoint (5), 2 houses reappoint (6), governor and commission reappoint (7), commission reappoints (8), life tenure (9). Because retention procedures change infrequently, judicial independence is averaged over 10-year periods. Switcher states changed at least once during 1866–2000. Log of duration of civil law is the average of the upper and lower bounds for duration of civil law reported in table 2.4.

The second column shows that the relationship is very similar if attention is restricted to states that actually changed their retention system at some point in the period 1866–2000. The third column shows that the relationship between political competition and partisan elections differed in common-law and civil-law states. The difference was, however, modest and not statistically significant. The fourth column yields similar results.

Table 5.7 shows that civil-law states increased judicial independence during periods of rising political competition, whereas common-law states did not. The first two columns are consistent with the results for partisan elections. The third and fourth columns suggest that the effect is primarily due to civil-law states. Common-law states did not generally increase independence during periods of rising political competition.

Judicial Terms

Longer tenure gives judges greater independence because they face elections or other reappointment procedures less often.[23] Depending on judges' age at the time of appointment to the state supreme court, if the judicial term is long enough, they may never face reappointment. Like judicial retention systems, there have been historical trends in term length.[24] Up until the 1830s high-level judges in most states had life tenure. During the 1840s and 1850s, average term length fell quite dramatically from about twenty years to about ten years, as many states shifted from life tenure to fixed terms. Average term length fell by another year during the remainder of the nineteenth century and averaged about nine years during the twentieth century. This section examines the factors that determined term lengths.

Figure 5.7 suggests that political competition has been related to term length over the period 1866–2000. The differences in levels of political competition in states with terms of less than ten years and states with terms greater than or equal to ten years were around twenty points for much of the period. The difference disappeared around 1980.

Figure 5.8 shows that civil-law states had consistently shorter terms than common-law states. The difference in term lengths was almost three years in

[23] Choi, Gulati, and Posner (2008) argue that while it increases independence, length of term does not necessarily increase judicial quality.

[24] See Epstein, Knight, and Shvetsova (2002). We follow them in assuming life tenure on average is twenty-five years.

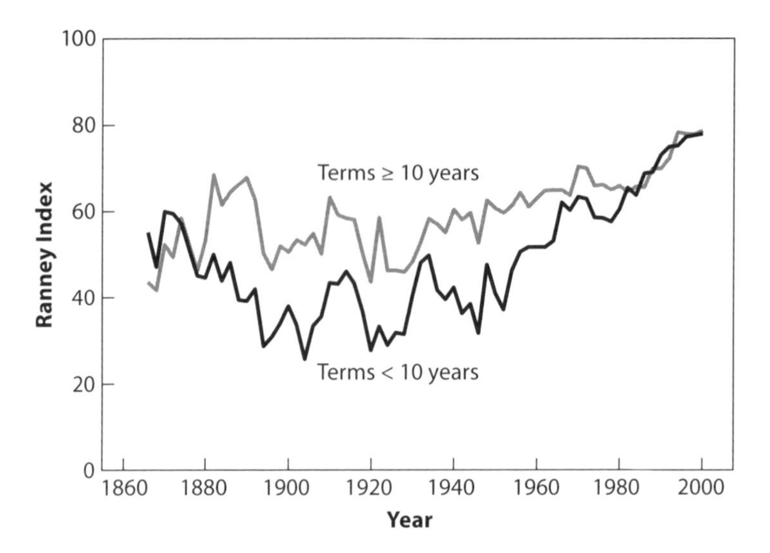

Figure 5.7. Political Competition and Term Length, 1866–2000.

Tenured judges are assigned 25-year terms. Louisiana is excluded because it retained civil law after entering the union, and Nebraska is dropped because it has a unicameral legislature for most of 1866–2000. This leaves 46 states in the sample. *Sources*: See the appendix to this chapter.

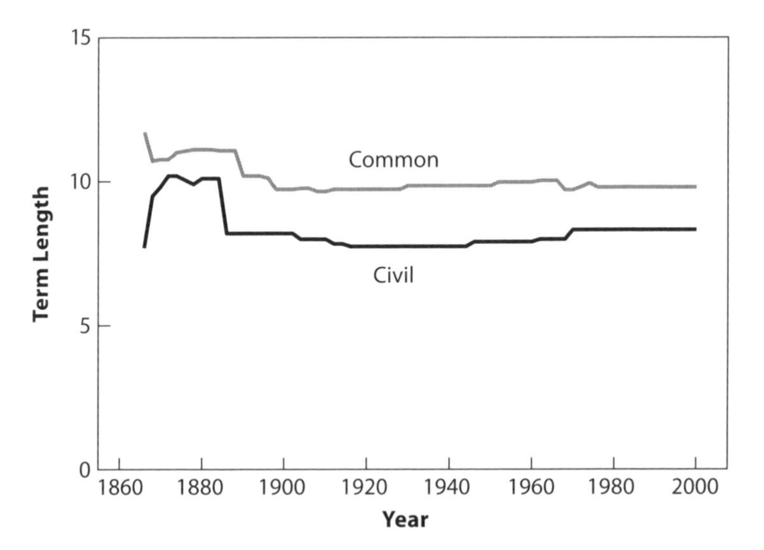

Figure 5.8. Term Length, 1866–2000.

Tenured judges are assigned 25-year terms. Louisiana is excluded because it retained civil law after entering the union, and Nebraska is dropped because it has a unicameral legislature for most of 1866–2000. This leaves 46 states in the sample. *Sources*: See the appendix to this chapter.

the early twentieth century and narrowed to about two years during the late twentieth century.

Table 5.8 shows that term length was positively related to political competition. The coefficient on the Ranney index is positive, statistically significant, and large in three of the four columns. For example, in the first column a twenty-point increase in the Ranney index would increase term length by 0.5 years. In the third and fourth columns, a twenty-point increase in the Ranney index would increase term length by 1.1 years. The civil variable is uniformly negative but not statistically significant.

Table 5.9 documents that civil-law states increased term length at higher levels of political competition than common-law states. The differences were large. For term length, the average Ranney index was 56 in common-law states and 86 in civil-law states. The decreases in term length in civil-law states occurred in southern states after the Civil War. Thus, the levels of political competition were very low.

Table 5.10 indicates that term lengths were only weakly related to increases in political competition.

Intermediate Appellate Courts

Intermediate appellate courts were introduced to allow the review of more cases, given the limited capacity of the state supreme court. Two states—Illinois and Missouri—adopted intermediate courts the 1870s. The pace of adoption was slow, however. In 1948, only eleven states had intermediate appellate courts. As late as 1956, only three more states had intermediate appellate courts. Adoption accelerated in the 1960s, and by 2000, intermediate appellate courts were operating most states.[25] Kagan et al. (1978) argued that population growth was a major reason for the emergence of these courts. Population growth created a case overload in state supreme courts. Creating intermediate appellate courts allowed state high courts to control their docket and make careful rulings.

Why would colonial legal origins matter for intermediate appellate courts? Legislatures in civil-law states may have pushed for the adoption of intermediate appellate courts because it improved their ability to control lower courts. Lower courts are difficult to control because the numbers of judges are large and their locations are spread throughout the state. With little prospect for judicial review, these courts have considerable autonomy. By expanding the

[25] For more on the modern role of intermediate appellate courts, see Chapp and Hanson (1990).

TABLE 5.8.
CIVIL LAW, POLITICAL COMPETITION, AND TERM LENGTHS

Dependent variable	*Term lengths 1866–2000*	*Term lengths 1866–2000*	*Term lengths 1866–2000*	*Term lengths 1866–2000*
Civil	−1.515	−1.292	−0.820	
	(1.180)	(1.242)	(1.227)	
Log of duration of civil law				−0.175
				(0.275)
Ranney	0.0238**	0.0249	0.0534**	0.0531**
	(0.0112)	(0.0200)	(0.0228)	(0.0228)
Pre-1847		3.310*	0.217	0.232
		(1.916)	(1.373)	(1.369)
Log of slavery		−0.310	0.483	0.483
		(0.631)	(0.477)	(0.477)
State FE	N	N	N	N
Decadal FE	N	Y	Y	Y
Sample	Full	Full	Switchers	Switchers
States	46	46	24	24
Observations	607	607	323	323
R^2	0.036	0.128	0.106	0.106

Notes: Standard errors are bootstrapped (500 repetitions) and clustered at the state level. The notations * and ** denote significance at the 10 percent and 5 percent levels. The constant is estimated but not reported. Nebraska is always dropped from the sample because it has a unicameral legislature. Louisiana is always dropped because it retained the civil law after joining the union. The constant is estimated but not reported. Term lengths run from 0 to 25 (lifetime tenure). Because term lengths change infrequently, term length is averaged over 10-year periods. Switcher states changed at least once during 1866–2000. Log of duration of civil law is the average of the upper and lower bounds for duration of civil law reported in table 2.4.

TABLE 5.9.
RANNEY INDEX AT THE TIME OF CHANGES IN TERM LENGTH AND POPULATION AT THE TIME OF ADOPTION OF INTERMEDIATE APPELLATE COURTS

	Common states	*Civil states*	*Difference*
Increasing term lengths	52.9 (4.34)	73.7 (3.80)	20.7*** (7.03)
Observations/states	66/12	29/6	95/18
Decreasing term lengths	67.8 (5.12)	18.7 (7.63)	–49.2*** (8.87)
Observations/states	27/6	16/4	43/10
Pop. when adopting int. appellate courts	3890.0 (207.2)	2882.4 (286.3)	−1007.6*** (366.6)
Observations/states	115/23	50/10	165/33
Removing intermediate appellate courts		No observations	

Notes: The Ranney index and population are measured at time of reform, two and four years before and after reform. Thus, each state that makes a reform has at most five observations. Term limit reforms are never reversed, and intermediate appellate court reforms are also never reversed. Each cell reports the mean, standard error (in parenthesis), and number of observations / number of states. The column for difference in means reports a test for the null that the difference in means is zero against a two-sided alternative. The notation *** denotes significance at the 1 percent level. The test statistic allows the common- and civil-law states to have unequal variances. Several states change partisan election, judicial independence, and term lengths more than once. As a result, the number of states is not simply the number of observations divided by five.

capacity for oversight in the form of intermediate appellate courts, the legislature can limit this autonomy. Further, since appellate courts are often located in the state capital, where the legislature and supreme court are also located, state legislators can more easily exert pressure on intermediate appellate judges than on the more numerous and geographically dispersed lower-court judges.

TABLE 5.10.
POLITICAL COMPETITION, CIVIL LAW, AND TERM LENGTHS WITH STATE FIXED EFFECTS

Dependent variable	*Term lengths 1866–2000*	*Term lengths 1866–2000*	*Term lengths 1866–2000*	*Term lengths 1866–2000*
Ranney	0.0046 (0.0092)	0.0085 (0.0175)	−0.0046 (0.0234)	−0.0053 (0.0234)
Ranney × Civil			0.0373 (0.0317)	
Log of duration of civil law				0.0084 (0.0070)
State FE	Y	Y	Y	Y
Decadal FE	N	N	N	N
Sample	Full	Switchers	Switchers	Switchers
States	46	24	24	24
Observations	607	323	323	323
R^2	0.900	0.678	0.687	0.688

Notes: Standard errors are bootstrapped (500 repetitions) and clustered at the state level. The constant is estimated but not reported. Nebraska is always dropped from the sample because it has a unicameral legislature. Louisiana is always dropped because it retained the civil law after joining the union. The constant is estimated but not reported. Term lengths run from 0 to 25 (lifetime tenure). Because term lengths change infrequently, term length is averaged over 10-year periods. Switcher states changed at least once during 1866–2000. Log of duration of civil law is the average of the upper and lower bounds for duration of civil law reported in table 2.4.

Figure 5.9 illustrates the strong association between population and intermediate appellate courts. The states with intermediate appellate courts were more heavily populated. Moreover, this difference in population between states that had and did not have intermediate appellate courts generally increased over time. In contrast, figure 5.10 suggests that political competition has played little role in the evolution of intermediate appellate courts.

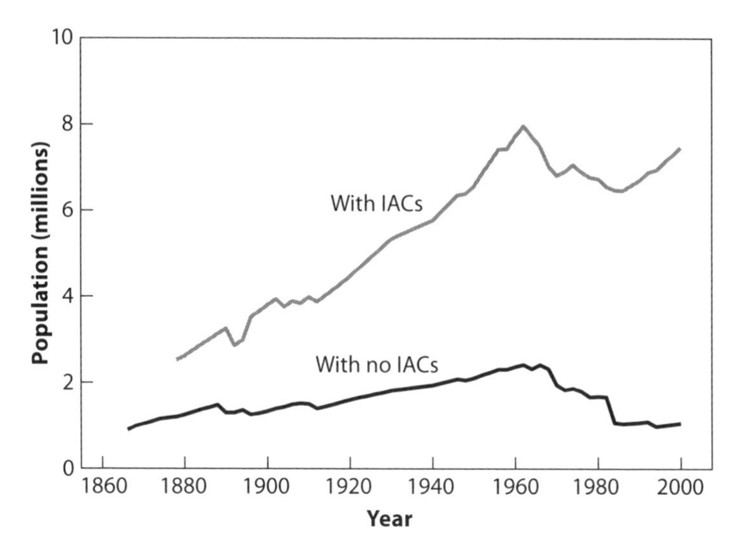

Figure 5.9. Population and Intermediate Appellate Courts, 1866–2000.

Louisiana is excluded because it retained civil law after entering the union, and Nebraska is dropped because it has a unicameral legislature for most of 1866–2000. This leaves 46 states in the sample. *Sources*: See the appendix to this chapter.

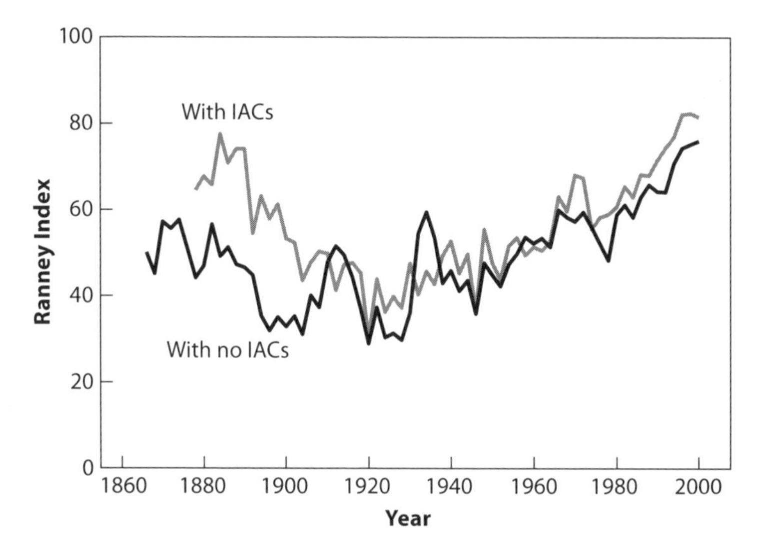

Figure 5.10. Political Competition and Intermediate Appellate Courts, 1866–2000.

Louisiana is excluded because it retained civil law after entering the union, and Nebraska is dropped because it has a unicameral legislature for most of 1866–2000. This leaves 46 states in the sample. *Sources*: See the appendix to this chapter.

Figure 5.11 indicates that civil-law states adopted intermediate appellate courts more rapidly than their common-law counterparts. By 1920, six of the twelve civil-law states in our sample had intermediate appellate courts. By 2000, all of them had intermediate appellate courts. In comparison, four of the thirty-four common-law states had intermediate appellate courts in 1920. By 2000, twenty-three of the common-law states had had intermediate appellate courts.

Table 5.11 confirms the importance of both population and civil law for intermediate appellate courts. The coefficients on both variables were statistically significant in all three specifications. The magnitudes of the effects were similar across all three specifications. Civil-law states' use of intermediate appellate courts was 0.22–0.23 higher than common-law states'. The coefficient on Log of population ranged from 0.16 to 0.21. Table 5.9 indicates that civil-law states adopted intermediate appellate courts at lower levels of population than common-law states.

Table 5.12 shows that states adopted intermediate appellate courts during periods of rising population within states. The differences between civil-law and common-law states in how they responded to increases in population were tiny and not statistically significant.

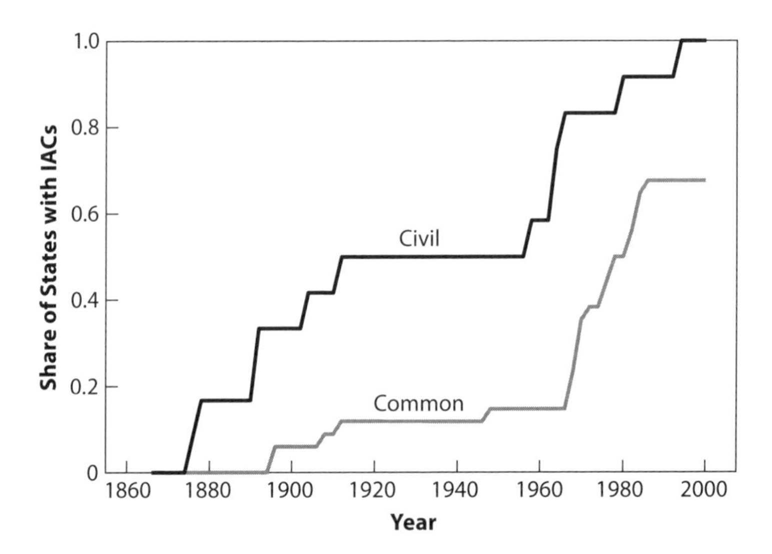

Figure 5.11. Use of Intermediate Appellate Courts, 1866–2000.

The vertical axis is the share of states that have operating intermediate appellate courts. Louisiana is excluded because it retained civil law after entering the union, and Nebraska is dropped because it has a unicameral legislature for most of 1866–2000. This leaves 46 states in the sample. *Sources*: See the appendix to this chapter.

TABLE 5.11.
CIVIL LAW, POLITICAL COMPETITION, AND INTERMEDIATE APPELLATE COURTS

Dependent variable	*IACs 1866–2000*	*IACs 1866–2000*	*IACs 1866–2000*	*IACs 1866–2000*
Civil	0.218*** (0.0714)	0.230*** (0.0797)	0.216*** (0.0789)	
Log of duration of civil law				0.0449*** (0.0173)
Ranney	0.0024** (0.0010)	0.0013 (0.0011)	0.0017 (0.0015)	0.0018 (0.0015)
Log of Population	0.206*** (0.0293)	0.163*** (0.0291)	0.178*** (0.0443)	0.183*** (0.0447)
Log of slavery		0.00565 (0.0253)	0.0132 (0.0293)	0.0145 (0.0294)
Pre-1847		0.0146 (0.0601)	–0.0012 (0.0963)	–0.0071 (0.0972)
State FE	N	N	N	N
Decadal FE	N	Y	Y	Y
Sample	Full	Full	Switchers	Switchers
States	46	46	35	35
Observations	605	605	464	464
R^2	0.439	0.527	0.565	0.561

Notes: Standard errors are bootstrapped (500 repetitions) and clustered at the state level. The notations ** and *** denote significance at the 5 percent and 1 percent level. The constant is estimated but not reported. Nebraska is always dropped from the sample because it has a unicameral legislature. Louisiana is always dropped because it retained the civil law after joining the union. The constant is estimated but not reported. Intermediate appellate courts equal 1 when they are operational and 0 otherwise. Log of population is from the decadal census and is interpolated from intracensus years. Intermediate appellate courts are averaged over 10-year periods. Switcher states changed at least once during 1866–2000. Log of duration of civil law is the average of the upper and lower bounds for duration of civil law reported in table 2.4.

TABLE 5.12.
CIVIL LAW, POLITICAL COMPETITION, AND INTERMEDIATE APPELLATE COURTS WITH FIXED EFFECTS

Dependent variable	*IACs 1866–2000*	*IACs 1866–2000*	*IACs 1866–2000*	*IACs 1866–2000*
Log of population	0.293*** (0.0366)	0.347*** (0.0361)	0.355*** (0.0491)	0.357*** (0.0490)
Log of population × civil law			–0.0164 (0.0661)	
Log of population × duration of civil law				–0.00463 (0.0139)
Ranney	0.0012 (0.0010)	0.0020* (0.0012)	0.0020 (0.0012)	0.0020 (0.0012)
State FE	Y	Y	Y	Y
Decadal FE	N	N	N	N
Sample	Full	Switchers	Switchers	Switchers
States	46	35	35	35
Observations	605	464	464	464
R^2	0.615	0.613	0.613	0.613

Notes: Standard errors are bootstrapped (500 repetitions) and clustered at the state level. The notations * and *** denote significance at the 10 percent and 1 percent levels. The constant is estimated but not reported. Nebraska is always dropped from the sample because it has a unicameral legislature. Louisiana is always dropped because it retained the civil law after joining the union. The constant is estimated but not reported. Intermediate appellate courts equal 1 when they are operational and 0 otherwise. Log of population is from the decadal census and is interpolated from intracensus years. Intermediate appellate courts are averaged over 10-year periods. Switcher states changed at least once during 1866–2000. Log of duration of civil law is the average of the upper and lower bounds for duration of civil law reported in table 2.4.

Conclusion

This chapter presented a model of how state legislatures choose judicial retention systems. The model generated predictions regarding the threshold levels of state political competition at which legislatures would change judicial retention procedures. This threshold level will be lower for common-law states than for civil-law states.

Tables 5.3–5.7 documented a number of differences between civil-law and common-law states in judicial retention procedures that were consistent with the model. Controlling for levels of political competition, the two groups of states differed in their use of partisan elections and in their levels of judicial independence. Civil-law states were more likely to use partisan elections and had lower levels of judicial independence than common-law states. Civil-law states made changes to judicial retention procedures at higher threshold levels of political competition than common-law states. As predicted by the model, states made changes during periods of rising political competition.

Other differences not directly related to the model were also documented. Two differences—one related to judicial tenure and one related to intermediate appellate courts—were of particular interest. Tenure of state high court judges was strongly related to state political competition. Civil-law states changed tenure at higher levels of political competition. Controlling for population and political competition, civil-law states adopted intermediate appellate courts earlier than common-law states. This is consistent with politicians in these states having a greater desire to monitor lower court judges than politicians in common-law states.

Appendix: Data Description and Sources

Judicial retention procedures, Epstein et al. index of judicial independence and length of judicial terms: These are taken from Epstein, Knight, and Shvetsova (2002). For the article and the data see http://epstein.law.northwestern.edu/research/selection.pdf and http://epstein.law.northwestern.edu/research/selection.por.

Intermediate appellate courts: The data are from Hanson (2001).

Decadal state population: The data are from the decennial censuses. Off-census years are interpolated.

Ranney index (political competition): The Ranney index is constructed from Burnham et al.'s (1986) data for the years 1834–1985 and from data collected by Tim Besley from the *Book of the States* for the years 1950–2000. For states in every election year, it lists the number of seats in the upper and lower house and the number of seats held by each party within each house. Most states elect parts of their state legislature either every year or every other year. A few states elect their state legislature every fourth year. The same data set is used to compute seats in the upper and lower house of each state legislature.

TABLE 5.1A.
PANEL REGRESSIONS

Dependent Variable	*Partisan elections*	*Judicial independence*	*Term lengths*	*Intermediate appellate courts*
	Cross-section (1900–1908)			
Civil	0.296*	−1.772*	−1.091	0.365**
	(0.156)	(0.885)	(1.514)	(0.137)
Ranney	0.0038	−0.0237*	0.0207	0.0044**
	(0.0025)	(0.0135)	(0.0267)	(0.0021)
Pre-1947	Y	Y	Y	Y
Log of slavery	Y	Y	Y	Y
Observations	44	44	44	46
	Panel analysis (base period is 1900–1908)			
Civil × 1866–1868	−0.0471	0.576	1.239	−0.303*
	(0.121)	(0.712)	(1.587)	(0.156)
Civil × 1870–1878	0.0545	−0.0942	0.740	−0.141
	(0.0879)	(0.515)	(1.031)	(0.139)
Civil × 1880–1888	0.0426	−0.274	−0.0540	−0.0364
	(0.0489)	(0.237)	(0.322)	(0.0690)
Civil × 1890–1898	0.0240	−0.0409	−0.522	−0.356**
	(0.145)	(0.766)	(1.238)	(0.178)
Civil × 1910–1918	0.0551	−0.141	−0.149	0.0315
	(0.0874)	(0.192)	(0.212)	(0.0769)
Civil × 1920–1928	0.169	−0.339	−0.201	0.0424
	(0.119)	(0.312)	(0.252)	(0.0857)
Civil × 1930–1938	0.120	−0.242	−0.311	0.0416
	(0.114)	(0.333)	(0.254)	(0.0877)

TABLE 5.1A. (*continued*)

Dependent variable	*Partisan elections*	*Judicial independence*	*Term lengths*	*Intermediate appellate courts*
Civil × 1940–1948	–0.0137	0.0145	–0.247	0.0391
	(0.170)	(0.414)	(0.275)	(0.0881)
Civil × 1950–1958	0.0231	–0.0240	–0.246	0.0384
	(0.171)	(0.410)	(0.311)	(0.0897)
Civil × 1960–1968	–0.0047	0.0539	–0.176	0.225*
	(0.169)	(0.424)	(0.341)	(0.117)
Civil × 1970–1978	–0.0875	0.369	0.380	0.116
	(0.194)	(0.521)	(0.696)	(0.167)
Civil × 1980–1988	–0.0709	0.355	0.466	–0.0266
	(0.194)	(0.528)	(0.732)	(0.167)
Civil × 1990–2000	–0.128	0.469	0.468	–0.0705
	(0.190)	(0.523)	(0.742)	(0.164)
State FE	Y	Y	Y	Y
Decadal FE	Y	Y	Y	Y
Decadal FE × pre-1847	Y	Y	Y	Y
Decadal FE × log of slavery	Y	Y	Y	Y
Ranney	Y	Y	Y	Y
Number of states	46	46	46	46
Observations	607	607	607	614

Notes: Standard errors are bootstrapped (500 repetitions) and clustered at the state level. The notations * and ** denote significance at the 10 percent and 5 percent levels. The constant is estimated but not reported. Louisiana and Nebraska are dropped for reasons already discussed in the text.

CHAPTER 6

Legislatures and Courts

THIS CHAPTER EXAMINES how colonial legal origin and political competition have shaped the funding of state courts. State judicial budgets are set by the state executive and legislative branches as part of a regular state budget exercise. Judicial budgets are used to pay for staff, facilities, and other resources such as professional experts who can evaluate complex cases, as well as judicial salaries. Fisher (1998) and Rosenburg (1991) have argued that the powers that governors and legislators have to shape judicial budgets serve as reasonable and useful checks on the judicial branch. However, governors and state legislatures may also use their power in setting judicial budgets to inappropriately influence judicial rulings.[1]

If colonial legal origins are influential, there may be systematic differences in judicial expenditures in common-law and civil-law states. Differences in expenditures may have implications for economic and social outcomes if restricted budgets lead to delay in processing criminals or in resolving contract, property, employment, or inheritance disputes. Civil-law states may provide fewer budgetary resources to their courts if funding is related to judicial independence. The state legislature may have more influence over an underfunded than a more fully funded judiciary. Further, following reforms that promote judicial independence, civil-law states may make different changes to expenditures than common-law states.

Using data on judicial expenditures for 1952–2000, we document differences between civil-law and common-law states in the levels at which the state funds courts and in how expenditures change following changes in retention procedures. The differences are substantial and appear to be driven by differences in court structure. One of the most important changes in retention procedures during this period was the move away from election procedures. Common-law states responded to this change by increasing expenditures, while civil-law states made smaller increases or even decreases in expenditures. One possible explanation is that civil-law states were using budgets to offset or at least not

[1] See Douglas and Hartley (2003).

increase the independence of the judiciary further. These patterns hold even when controls for slavery and entry into the union before 1847 are included.

Judicial Expenditures

Data on state judicial and legal expenditures that are comparable across states are available beginning in the early 1950s. The data come from the Census of Governments and include all state expenditures on criminal and civil courts.[2] These expenditures include salaries for judges and court reporters and payments for witness fees, as well as payments to legal departments, general counsels, solicitors, and prosecuting and district attorneys. In 1982, this variable began to include payments for public defenders and legal services.

One thing to note is that state expenditure does not include local expenditure.[3] Local funds pay for municipal and county courts, which have limited jurisdiction. They commonly handle traffic, ordinance, misdemeanor, and smaller civil cases. Data on local expenditures begins in 1977. State money typically supports general jurisdiction courts. These courts include the supreme court, intermediate appellate courts, and district or circuit courts. Because our interest is in these courts and because data on state expenditures are available for a much longer period, our focus is on state expenditures.

Tables 6.1 and 6.2 provide summary statistics and correlations for the main variables of interest. Expenditures were positively correlated with judicial independence and negatively correlated with the use of partisan elections, any type of election, and civil law. Not surprisingly, the two expenditure variables were fairly highly correlated. Salaries were weakly correlated with all of the variables.

Figure 6.1 indicates that per capita expenditures have been fairly persistent over time. States with above average expenditures in 1952–1958 had above average expenditures in other periods.

Figure 6.2 graphs the log of (real) judicial expenditures per capita in common-law and civil-law states over the period 1952–2000. Judicial expenditures per capita grew rapidly in real terms, in part because of increasing urbanization

[2] Data on state and local judicial budgets are taken from the U.S. Census Bureau, Annual Survey of State and Local Government Finances and Census of Governments, various years. We thank John L. Curry of the Bureau of the Census for supplying these data and for helping us interpret them. See U.S. Census Bureau (2001).

[3] On the evolution of local versus state control and funding of courts, see Tobin (1999).

Table 6.1.
Summary Statistics for Judicial Retention and Expenditures

	Judicial independence	*Elections*	*Partisan*	*Deflated per capita judicial expenditures*	*Judicial expenditures shares*	*High court salaries*
Average	3.07	0.53	0.23	16.86	0.68	99,011
Standard deviation	2.22	0.50	0.42	19.51	0.51	20,560
Minimum	1	0	0	0.78	0.08	48,194
Maximum	9	1	1	116.69	2.79	179,006
Number of states that change	21 6 civil 15 com	15 5 civil 10 com	13 5 civil 8 com	46	46	46

Sources: See the appendix to this chapter.

Notes: Louisiana is excluded because it retained the civil law after entering the union and Nebraska is dropped because it has a unicameral legislature for most of 1866–2000. This leaves 46 states in the sample.

Table 6.2.
Correlations for Judicial Retention and Expenditures

	Judicial independence	*Elections*	*Partisan*	*Deflated per capita judicial expenditures*	*Judicial expenditures shares*	*High court salaries*
Judicial independence	1.00					
Elections	–0.73	1.00				
Partisan	–0.52	0.52	1.00			
Per capita expenditures	0.29	–0.35	–0.23	1.00		
Expenditures shares	0.35	–0.41	–0.20	0.90	1.00	
High court salaries	0.10	–0.20	–0.07	0.20	0.14	1.00
Civil	–0.30	0.08	0.32	–0.15	–0.16	0.16

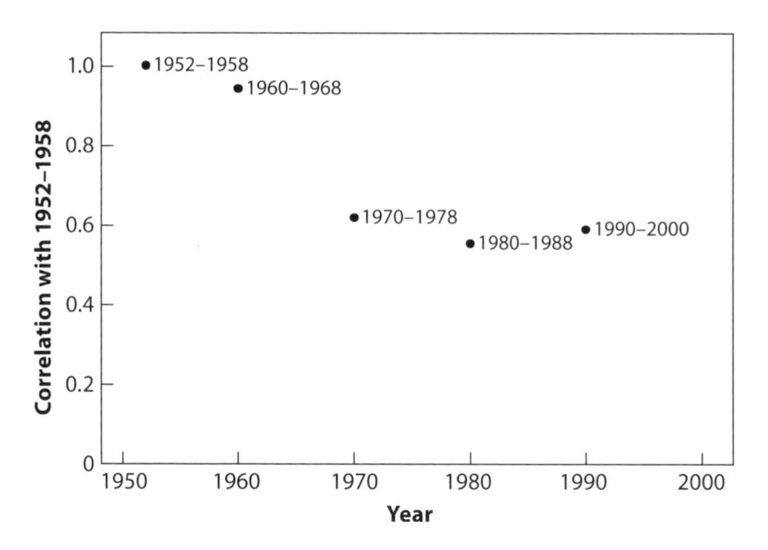

Figure 6.1. Persistence of Deflated Judicial Expenditures Per Capita, 1952–2000.

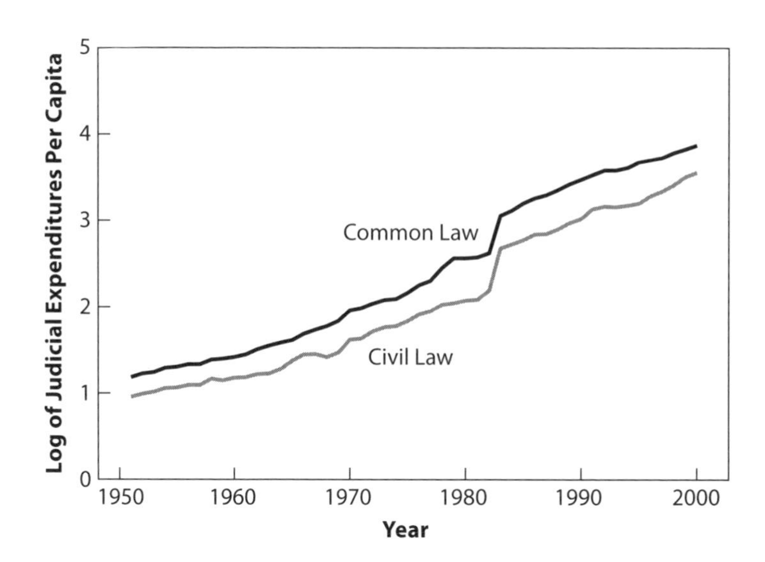

Figure 6.2. Deflated Judicial Expenditures Per Capita, 1952–2000.

Louisiana is excluded because it retained civil law after entering the union, and Nebraska is dropped because it has a unicameral legislature for most of 1866–2000. This leaves 46 states in the sample. Expenditures are deflated. The base year is 2000. *Sources*: See the appendix to this chapter.

and rising incomes. The sharp increase in the expenditures of both types of states in the early 1980s reflects the inclusion of payments for public defenders and legal services. Over the entire period, judicial budgets per capita were 35 percent lower in civil-law states than in common-law states. In 2000 an average civil-law state spent $37.52 per capita and an average common-law state spent $52.31.

Figure 6.3 shows that the pattern is somewhat similar for judicial expenditures as a percentage of state budgets. Judicial expenditure shares were relatively constant through the mid-1970s and then began to grow. This suggests that the rising per capita judicial expenditures in figure 6.2 before the mid-1970s may have been a result of expansion in state income. After the mid-1970s, judicial expenditures rose relative to total expenditure. Over the whole period, judicial expenditure shares were 25 percent lower in the civil-law states.

Figure 6.4 plots the log of (real) supreme-court salaries in common-law and civil-law states for the period 1960–2000. Surprisingly, in light of the previous two figures, salaries were 8 percent higher in civil-law states.

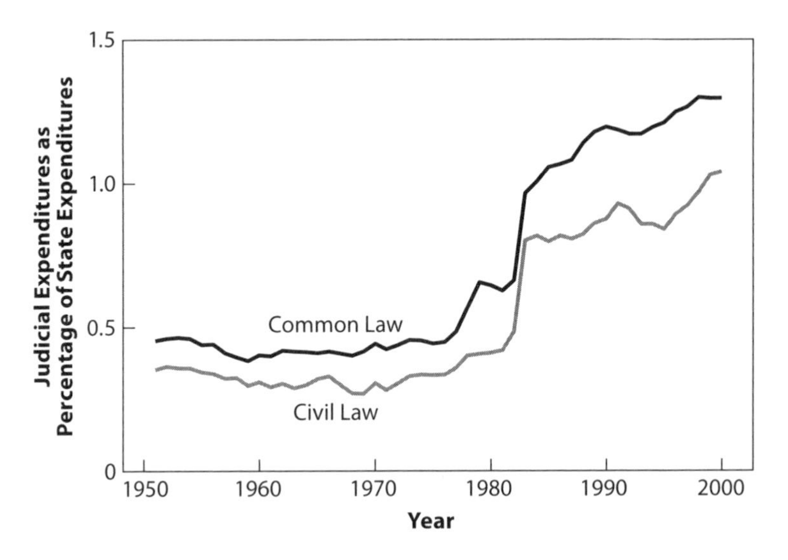

Figure 6.3. Judicial Expenditures as Percentage of State Expenditure, 1952–2000.

Louisiana is excluded because it retained civil law after entering the union, and Nebraska is dropped because it has a unicameral legislature for most of 1866–2000. This leaves 46 states in the sample. *Sources*: See the appendix to this chapter.

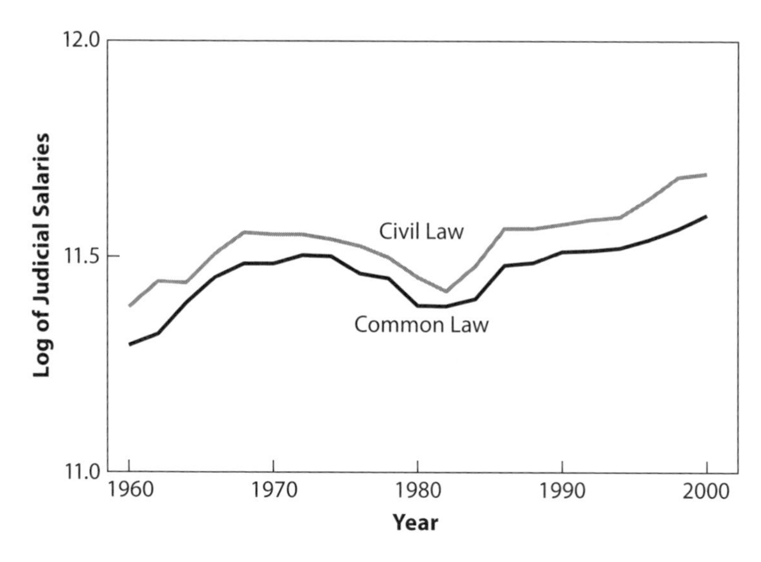

Figure 6.4. State Supreme Court Salaries, 1960–2000.

Louisiana is excluded because it retained civil law after entering the union, and Nebraska is dropped because it has a unicameral legislature for most of 1866–2000. This leaves 46 states in the sample. Salaries are deflated. The base year is 2000. *Sources*: See the appendix to this chapter.

Table 6.3 explores these patterns further in a regression framework. It examines the effect of civil law, entry into the union before 1847, the log of the share of slaves in 1860, and political competition on judicial expenditures, judicial expenditure shares, and judicial salaries. Consistent with figures 6.2–6.4, table 6.3 shows that civil-law states had lower expenditures and higher salaries than common-law states. In addition, states that entered the union before 1847 generally had higher budgets and higher salaries than states that entered after 1847. Slavery and political competition had limited impacts on expenditures and salaries. Political competition was only statistically significant for salaries, and the effect was fairly modest. A twenty-point increase in political competition would have raised salaries by about 3.8 percent.[4]

[4] In unreported regressions in which the sample was restricted to states that made some change to judicial independence, the point estimates for the effect of civil were larger, but the small sample size (twenty-one states) meant that the differences were no longer statistically significant.

TABLE 6.3.
DETERMINANTS OF EXPENDITURE AND SALARIES

Dependent variable	*Log of per capita expenditures 1952–2000*	*Judicial share 1952–2000*	*Log of salaries 1960–2000*
Civil	−0.284*	−0.162	0.0857*
	(0.150)	(0.0974)	(0.0509)
Pre-1847	0.227*	0.223**	0.153***
	(0.129)	(0.0905)	(0.0473)
Log of slave share	−0.0878	−0.0424	−0.000357
	(0.0539)	(0.0398)	(0.0124)
Ranney	−0.0016	−0.0014	0.0019***
	(0.0018)	(0.0014)	(0.0006)
State FE	N	N	N
Year FE	Y	Y	Y
Sample	Full	Full	Full
States	46	46	46
Observations	1146	1146	959
R^2	0.776	0.486	0.932

Notes: Standard errors are robust and clustered at the state level. The notations *, **, and *** denote significance at the 10 percent, 5 percent, and 1 percent levels. The constant is estimated but not reported. Nebraska is always dropped from the sample because it has a unicameral legislature. Louisiana is always dropped because it retained the civil law after joining the union.

Why would civil-law states have lower expenditures? One possible reason might be that a small number of common-law states, such as Delaware, may be spending larger than average amounts, because they have invested in courts as part of an overall strategy for recruiting and retaining business. Delaware does spend a large amount on its courts. The distribution of expenditures on state

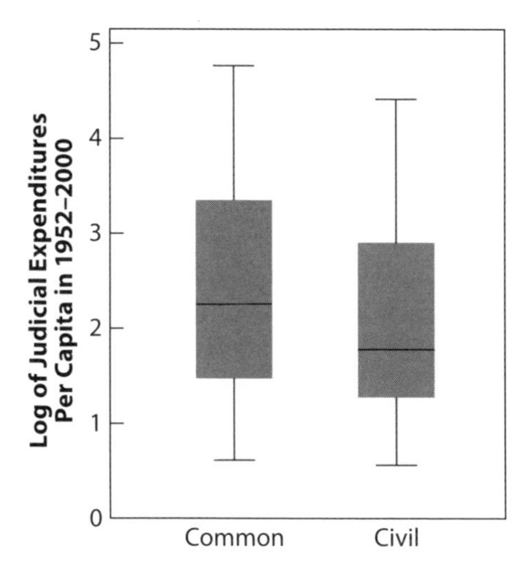

Figure 6.5. Judicial Expenditures Per Capita in Common-law and Civil-Law States, 1952–2000.

Louisiana is excluded because it retained civil law after entering the union, and Nebraska is dropped because it has a unicameral legislature for most of 1866–2000. This leaves 46 states in the sample. Expenditures are deflated. The base year is 2000. In each box, the central bar is the median state, the box contains the interquartile range, and the T-sticks contain the entire population when no state is more than two standard deviations above or below average. When a state is more than two standard deviations above or below the average, it is depicted with a black circle that lies outside the T-sticks. *Sources*: See the appendix to this chapter.

courts is not, however, being driven by a small number of outliers. Figures 6.5 and 6.6 plot the distribution of per capita expenditures in common-law and civil-law states overall and for one year, 1976, which was chosen because it fell in the middle of the sample. Some common-law states did spend large amounts on their courts, but the distributions differ at many other points, including the medians.

Table 6.4 adds variables related to the structure of state courts and population to the variables in table 6.3. The population allows for economies of scale in provision of court services. For example, state supreme courts and intermediate appellate courts were largely fixed costs, so their average costs would have declined with population. The coefficients on the initial conditions—civil, entry before 1847, and slavery—are no longer statistically significant. Their effect on expenditures appears to operate through their effect on judicial independence and intermediate appellate courts.

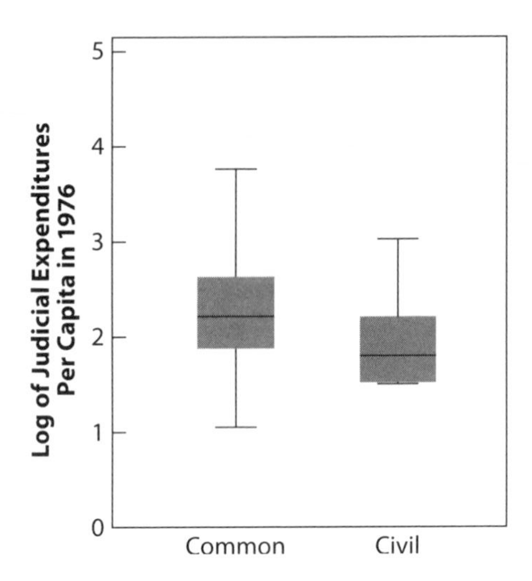

Figure 6.6. Judicial Expenditures Per Capita in Common-Law and Civil-Law States in 1976.

Louisiana is excluded because it retained civil law after entering the union, and Nebraska is dropped because it has a unicameral legislature for most of 1866–2000. This leaves 46 states in the sample. In each box, the central bar is the median state, the box contains the interquartile range, and the T-sticks contain the entire population when no state is more than two standard deviations above or below the average. When a state is more than two standard deviations above or below the average, it is depicted with a black circle that lies outside the T-sticks. Expenditures are deflated. The base year is 2000. *Sources*: See the appendix to this chapter.

The first two columns show that states with more independent high court judges, intermediate appellate courts, and smaller populations had higher expenditures. For example, in the first column a one-point increase in judicial independence was associated with a 6 percent increase in per capita expenditures. States with intermediate appellate courts spent about 20 percent more per capita than states without intermediate appellate courts. The negative and statistically significant coefficient on the log of population in shows that larger states spent less per capita. This likely reflects the presence of some economies of scale in the provision of court services.

Salaries for high court judges were higher in states with great judicial independence, in larger states, and in states with higher levels of political competition. Their positions may have been seen as relatively more important in these states. A one-point increase in judicial independence is associated with a 1.3 percent in salaries. This is similar in magnitude to the effect of a 10 percent

TABLE 6.4.
COURT STRUCTURE AND EXPENDITURES AND SALARIES

Dependent variable	*Log of per capita expenditures 1952–2000*	*Judicial share 1952–2000*	*Log of salaries 1960–2000*
Judicial independence	0.0585* (0.0332)	0.0562** (0.0222)	0.0129** (0.00556)
Intermediate appellate courts	0.213** (0.101)	0.153** (0.0753)	–0.0111 (0.0197)
Log of population	–0.213*** (0.0683)	–0.109*** (0.0400)	0.138*** (0.0184)
Civil	–0.140 (0.151)	–0.0754 (0.0946)	–0.0009 (0.0270)
Pre-1847	0.199 (0.138)	0.144 (0.0969)	0.0403 (0.0337)
Log of slave share	–0.0351 (0.0574)	–0.0014 (0.0429)	–0.0080 (0.0090)
Ranney	0.0002 (0.0014)	–0.0003 (0.0011)	0.0009* (0.0005)
State FE	N	N	N
Year FE	Y	Y	Y
Sample	Full	Full	Full
States	46	46	46
Observations	1146	1146	959
R^2	0.810	0.549	0.653

Notes: Standard errors are robust and clustered at the state level. The notations *, **, and *** denote significance at the 10 percent, 5 percent, and 1 percent levels. The constant is estimated but not reported. Nebraska is always dropped from the sample because it has a unicameral legislature. Louisiana is always dropped because it retained the civil law after joining the union.

increase in population, which was 1.4 percent. A twenty-point increase in the Ranney score would increase salaries by 1.8 percent.

Overall, the evidence suggests that civil-law states had lower expenditures on courts than common-law states and that these differences were largely related to the differences in court structures.

Changes in Judicial Retention and Judicial Expenditure

A state that changes judicial retention procedures may also chose to make changes to the level of judicial expenditure. Expenditures may rise if changes in retention procedures are part of a broader court improvement program. Expenditures could also fall. State legislators and governors may use cuts to offset increased judicial independence following the removal of elections. Thus, state expenditures following the removal of elections provide information about the preferences the governors and state legislatures have for an independent or weak judiciary. Over the period 1952–2000 of the forty-six states in the sample, twenty-one made some type of change to judicial independence; thirteen removed partisan elections; and fifteen removed elections.

Table 6.5 indicates that some changes in judicial retention procedures within a state were associated with changes in per capita expenditures. Changes in judicial independence did not appear to have an effect on expenditures, but changes in the use of elections had a statistically significant effect. The removal of elections, which includes both partisan and nonpartisan elections, was arguably the most important change that a state could make to retention systems. States that moved away from elections increased per capita expenditures by 18 percent. Removing partisan elections had a more limited effect, because a number of these states were moving from partisan elections to nonpartisan elections. Many commentators would argue that the net effect of this change on judicial independence is small.

The adoption of intermediate appellate courts and population growth were also associated with statistically significant changes in per capita expenditures. States that adopted intermediate appellate courts increased per capita expenditures by 19–21 percent. A 10 percent increase in population was associated with a 3–4 percent decrease in per capita spending. That may be the result of economies of scale in the provision of court services, or it may reflect the fact that faster-growth states are simply slow to increase court budgets.

Table 6.6 shows that the removal of elections influenced budget shares but not judicial salaries. The adoption of intermediate appellate courts and population were not statistically significantly associated with budget shares or salaries.

TABLE 6.5.
CHANGES IN PER CAPITA EXPENDITURES WITHIN STATES

Dependent variable	*Log of per capita expenditures 1952–2000*	*Log of per capita expenditures 1952–2000*	*Log of per capita expenditures 1952–2000*
Judicial independence	0.0533 (0.0425)		
Elections		–0.179* (0.105)	
Partisan			–0.0664 (0.138)
Intermediate appellate courts	0.208** (0.0870)	0.195** (0.0853)	0.194** (0.0882)
Log of population	–0.380* (0.198)	–0.392** (0.188)	–0.349* (0.205)
Ranney	–0.0009 (0.0009)	–0.0008 (0.0009)	–0.0010 (0.0009)
State FE	Y	Y	Y
Year FE	Y	Y	Y
Sample	Full	Full	Full
States	46	46	46
Observations	1146	1146	1146
R^2	0.933	0.933	0.932

Notes: Standard errors are robust and clustered at the state level. The notations * and ** denote significance at the 10 percent and 5 percent levels. The constant is estimated but not reported. Nebraska is always dropped from the sample because it has a unicameral legislature. Louisiana is always dropped because it retained the civil law after joining the union.

TABLE 6.6.
CHANGES IN EXPENDITURE SHARES AND SALARIES WITHIN STATES

Dependent variable	*Judicial share 1952–2000*	*Log of salaries 1960–2000*
Elections	–0.184*	0.0067
	(0.0928)	(0.0396)
Intermediate appellate courts	0.111	0.0124
	(0.0748)	(0.0247)
Log of population	–0.0633	0.133
	(0.177)	(0.0847)
Ranney	–0.0012	0.0007
	(0.0010)	(0.0004)
State FE	Y	Y
Year FE	Y	Y
Sample	Full	Full
States	46	46
Observations	1146	959
R^2	0.789	0.980

Notes: Standard errors are robust and clustered at the state level. The notation * denotes significance at the 10 percent level. The constant is estimated but not reported. Nebraska is always dropped from the sample because it has a unicameral legislature. Louisiana is always dropped because it retained the civil law after joining the union.

One question regarding the change in expenditure is how the money was spent. The evidence on salaries suggests that the money did not go towards increased salaries for existing judges. Unfortunately, detailed data on state expenditure are only available after 1977. Only four states switched after this period—Georgia, Mississippi, New Mexico, and Tennessee. As a result, we cannot say anything about how the judicial branch spent the additional money.

In interpreting the effects, another possible concern is that elements of expenditure have been omitted. Indeed, state expenditure represents less than half of total expenditure on the judiciary, because it typically only covers spending on courts of general jurisdiction, intermediate appellate courts, and the state supreme court.[5] Local governments maintain a wide variety of limited-jurisdiction courts including municipal and county courts that handle traffic issues, family matters, minor criminal cases, juvenile cases, small claims, and some civil matters.[6] The number of judicial officers associated with limited jurisdiction courts is greater than the number of judicial officers associated with unified courts and general jurisdiction courts, and they typically have higher caseloads. Unfortunately, data on local expenditures on these courts only become available in 1977.

For the four states that switched after 1977 there does not appear to have been any change in local spending as a result of changes in judicial retention procedures for the state supreme court. The lack of a change is not surprising, since any effect on local courts should be tiny. It does indicate that changes in retention procedures were not coupled with changes in the mix of state and local funding for courts.

Figures 6.7–6.10 show detrended state judicial budget shares before and after the removal of elections in the two civil-law states, Illinois and Indiana, and two common-law states, Colorado and New York. For purposes of comparison, all four states are nonslave (northern) states. To facilitate comparison across years, 0 represents the average share in a given year. Recall that the national average in 1952 was 0.4 percent and by 2000 it was 1.1 percent. In the full sample and most subperiods, 0.5 percent was roughly one standard deviation.

The figures show that increases in expenditure were generally smaller following reform in the civil-law states (Illinois and Indiana) than in the common-law states (Colorado and New York). Prior to reform, all four states were at or slightly below the national average. Shortly after the removal of partisan elections in Illinois, expenditure shares briefly increased to roughly 0.4 above average and then declined. In Indiana, expenditure shares remained constant around the time of reform. Colorado increased expenditure shares by more than 0.5 after reform. New York increased expenditure by about 1.5, although after a few years expenditures began to decline towards the mean. In sum, the change in expenditure following reform appears to have been bigger in the two common-law states than in the two civil-law states.

Table 6.7 explores the extent to which civil-law and common-law states differed in their responses to changes in retention.[7] One thing to keep in mind is

[5] Local courts are usually courts of limited jurisdiction. For information on court structure and funding by state, see National Center for State Courts (2008).

[6] Traffic and ordinance violation cases alone represent more than half of all incoming cases for state courts. For a wealth of information on state courts, see LaFountain et al. (2008).

[7] The results are very similar if the effect of intermediate appellate courts is also interacted with civil law.

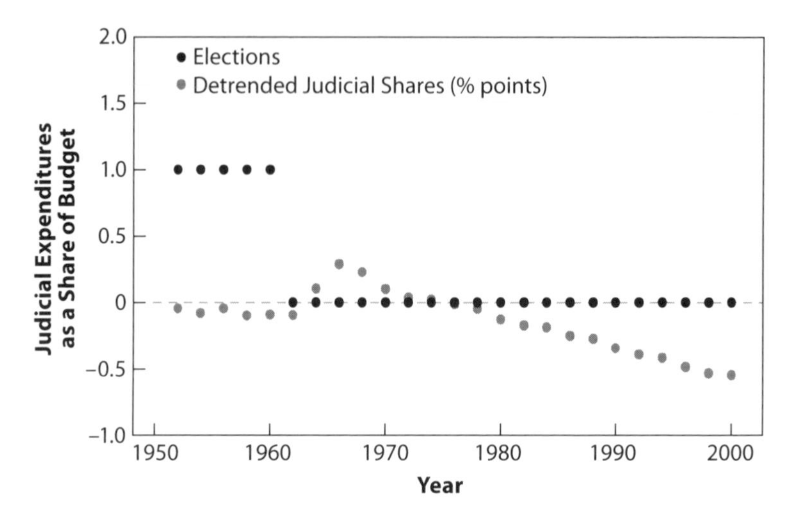

Figure 6.7. Judicial Retention and Detrended Expenditure Shares in Illinois (Civil).

Elections = 1 indicates that a state uses partisan or nonpartisan elections to retain state high court judges. Elections = 0 indicates that judges are reappointed or are subject to uncontested retention elections.

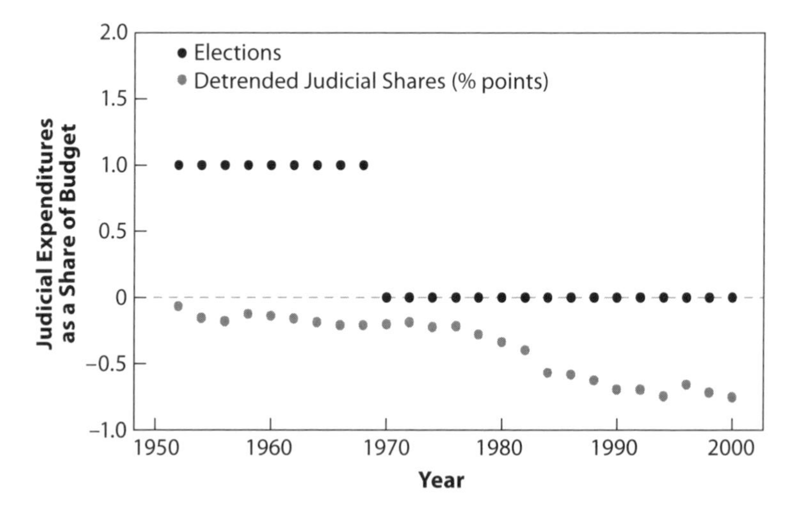

Figure 6.8. Judicial Retention and Detrended Expenditure Shares in Indiana (Civil).

Elections = 1 indicates that a state uses partisan or nonpartisan elections to retain state high court judges. Elections = 0 indicates that judges are reappointed or are subject to uncontested retention elections.

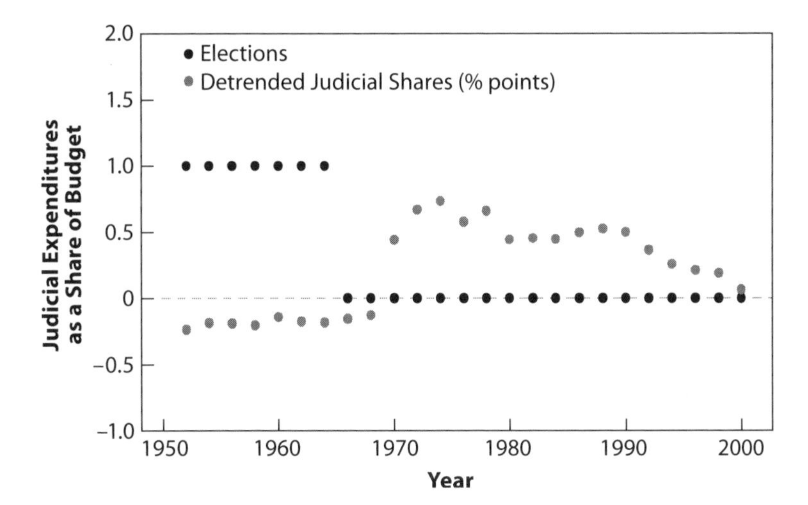

Figure 6.9. Judicial Retention and Detrended Expenditure Shares in Colorado (Common).

Elections = 1 indicates that a state uses partisan or nonpartisan elections to retain state high court judges. Elections = 0 indicates that judges are reappointed or are subject to uncontested retention elections.

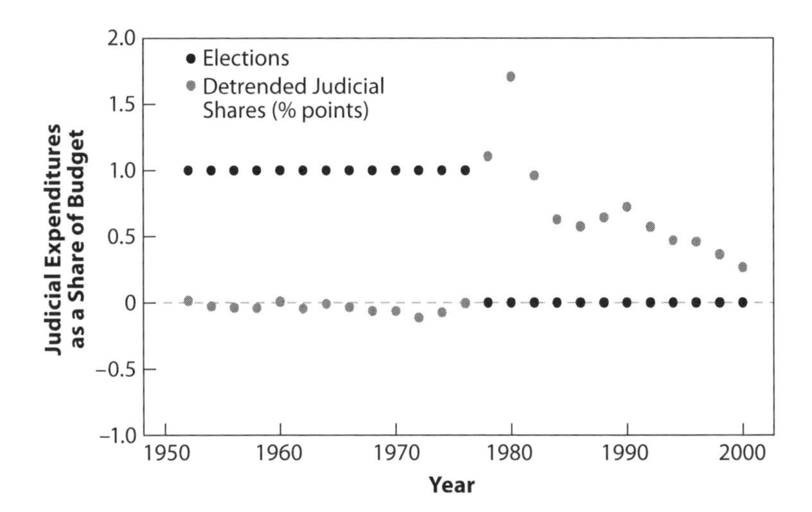

Figure 6.10. Judicial Retention and Detrended Expenditure Shares in New York (Common).

Elections = 1 indicates that a state uses partisan or nonpartisan elections to retain state high court judges. Elections = 0 indicates that judges are reappointed or are subject to uncontested retention elections.

Table 6.7.
Changes in Per Capita Expenditures in Common-Law and Civil-Law States

Dependent variable	*Log of per capita expenditures 1952–2000*	*Log of per capita expenditures 1952–2000*	*Log of per capita expenditures 1952–2000*
Elections	−0.228*	−0.125	−0.184*
	(0.132)	(0.0927)	(0.103)
Elections × Civil	0.234	0.125	0.111
	(0.152)	(0.116)	(0.109)
Intermediate appellate courts	0.175*	0.203**	0.286**
	(0.0871)	(0.0854)	(0.098)
Ranney	−0.0007	−0.0009	−0.0008
	(0.0010)	(0.0010)	(0.0010)
State FE	Y	Y	Y
Year FE	Y	Y	Y
Sample	Full	Full, drop NY	Change indep
States	46	45	21
Observations	1146	1121	498
R^2	0.932	0.933	0.944

Notes: Because of the small number of states in specification (3) standard errors in all three columns are bootstrapped (500 repetitions) and clustered at the state level. The notations * and ** denote significance at the 10 percent and 5 percent levels. The constant is estimated but not reported. Nebraska is always dropped from the sample because it has a unicameral legislature. Louisiana is always dropped because it retained the civil law after joining the union.

that the number of states that change is relatively small, and thus any differences between civil-law and common-law states are not likely to be statistically significant. The difference is very large in the first column. Common-law states that moved away from elections increased per capita expenditures by 23 percent. And civil-law states differed from common-law states by –23 percent. In other words, civil-law states did not increase budgets at all.

One thing to note is that New York was something of an outlier. The increases in per capita expenditures there were considerably larger than in other states. The second column in the table drops New York. Although the magnitudes are smaller, the patterns are similar to the first column. Common-law states that moved away from elections increased per capita expenditures by 13 percent. And civil-law states did not increase budgets at all. The third column drops New York and restricts the sample to states that made changes in judicial independence at some point during the period 1952–2000. Common-law states that moved away from elections increased per capita expenditures by 19 percent, while civil-law states increased budgets by 10 percent. Intermediate appellate courts are strongly associated with high expenditure.

One question is whether the differences in response to the move away from elections were due to colonial legal origin or some other factor. Table 6.8 adds additional interaction terms for pre-1847 entry into the union and for slavery. Addition of these terms drives down statistical significance, but does not change the basic patterns. Common-law states that moved away from elections tended to increase budgets. Depending on the specification, civil-law states on net either decreased budgets or increased them by much smaller amounts than common-law states.

Conclusion

This chapter and the previous chapter have documented the roles that political competition and colonial legal systems have played in shaping state court systems. State court systems exhibited persistence along a variety of dimensions including levels of judicial independence, the length of judges' terms, and levels of expenditure. The model both offered an explanation for why change would be infrequent and explained how political competition and colonial legal systems would be related to change when it did occur.

The data supported the argument that civil-law and common-law states had different balances of power between the legislature and the judiciary. Civil-law states were more likely to use partisan elections, had lower judicial independence, and increased judicial independence and term length at higher levels

TABLE 6.8.
CHANGES IN PER CAPITA EXPENDITURES IN COMMON-LAW AND CIVIL-LAW STATES WITH ADDITIONAL CONTROLS

Dependent variable	*Log of per capita expenditures 1952–2000*	*Log of per capita expenditures 1952–2000*	*Log of per capita expenditures 1952–2000*
Elections	–0.174	–0.132	–0.188
	(0.132)	(0.119)	(0.118)
Elections × Civil	0.257	0.110	0.0940
	(0.500)	(1.828)	(0.152)
Elections × Pre 1847	–0.213	0.0821	0.0915
	(0.750)	(1.817)	(1.191)
Elections × Log of slave share	0.0537	–0.0285	–0.0360
	(0.746)	(0.964)	(0.789)
Intermediate appellate courts	0.186**	0.201**	0.281***
	(0.0821)	(0.0823)	(0.100)
Ranney	–0.0008	–0.0009	–0.0008
	(0.0010)	(0.0010)	(0.0011)
State FE	Y	Y	Y
Year FE	Y	Y	Y
Sample	Full	Full, drop NY	Change indep
States	46	45	21
Observations	1146	1121	498
R^2	0.932	0.933	0.945

Notes: Because of the small number of states in specification (3) standard errors in all three columns are bootstrapped (500 repetitions) and clustered at the state level. The notations ** and *** denote significance at the 5 percent and 1 percent levels. The constant is estimated but not reported. Nebraska is always dropped from the sample because it has a unicameral legislature. Louisiana is always dropped because it retained the civil law after joining the union.

of political competition than common-law states. Civil-law states were more likely to have intermediate appellate courts and adopted them at lower levels of population than common-law states. Civil-law states also had lower state expenditures on courts than common-law states. This appears to have been due to differences in court structure. Finally, civil-law states appear to have responded differently than common-law states to changes in judicial retention. Following a move away from elections, common-law states increased expenditure, while civil-law states made smaller increases or even decreases in expenditure. Taken together with the evidence on the very high level of political competition at the time of changes in retention systems in civil-law states as compared to common-law states, these results suggest that the relevant political actors in civil-law states may have been less enamored with the change than the relevant political actors in common-law states.

The most significant concern with respect to civil law is simply that it is capturing the effect of some other variable. Inclusion of the two most obvious candidates—timing of entry, which determined the initial retention system, and slavery—as controls had limited effects on our main results. This does not rule out the possibility of yet another variable being the true cause of these patterns, but the variable would have to be correlated with colonial legal origins and explain the observed patterns. The preponderance of historical evidence points to the colonial legal system as having shaped the initial balance of power. This balance of power together with political competition shaped the structure of state courts.

The differences between common-law and civil-law state courts are not merely academic. The available evidence suggests that judicial retention systems, and possibly other aspects of state courts such as judicial budgets, affect judicial decision-making and that judicial decision-making shapes economic and social outcomes.

Appendix: Data Description and Sources

State and Local Expenditure: These are taken from the U.S. Census Bureau, Annual Survey of State and Local Government Finances and Census of Governments, various years. We thank John L. Curry of the Bureau of the Census for supplying this data.

Judicial Salary Data: These are taken from the *Book of the States*. Council of State Government (various years).

TABLE 6.1A.
REMOVAL OF ELECTIONS (PARTISAN AND NONPARTISAN), 1952–2000

State	*Civil*	*Year of reform*	*Eliminated election system*	*Intermediate appellate courts first established*
Arizona	1	1974	Nonpartisan	1964
Colorado	0	1966	Partisan	1970
Florida	1	1976	Nonpartisan	1957
Illinois	1	1962	Partisan	1877
Indiana	1	1970	Partisan	1891
Iowa	0	1962	Partisan	1976
Kansas	0	1958	Partisan	1977
Maryland	0	1976	Nonpartisan	1967
New Mexico	1	1988	Partisan	1965

TABLE 6.1A. (*continued*)

State	*Civil*	*Year of reform*	*Eliminated election system*	*Intermediate appellate courts first established*
New York	0	1977	Partisan	1895
Oklahoma	0	1967	Partisan	1968
South Dakota	0	1980	Nonpartisan	Never
Tennessee	0	1994	Nonpartisan	1895
Utah	0	1980	Partisan	1986
Wyoming	0	1972	Partisan	Never

CHAPTER 7

Institutions and Outcomes

This book has documented the persistence of state political and legal institutions, the mechanisms through which initial conditions acted on state political and legal institutions, and the reasons for persistence. The main reason that many scholars, including us, care about institutions and their persistence is because institutions affect outcomes.

Although many different outcomes might be of interest, for simplicity we focus on state per capita income. It represents an important measure of total economic activity in a society. Courts and legislatures may have other objectives, but arguably one of their functions is to maximize the size of the societal pie by offering incentives for economic activity.

The remainder of the chapter summarizes our main findings with respect to persistence and mechanisms, examines the effect of state political and legal institutions on per capita income, and discusses important areas for future research.

Persistence and Mechanisms: State Legislatures

Chapters 3 and 4 provided a framework within which to understand the evolution of American state legislatures. Various elements of our story have already been widely discussed in the political and historical literatures. While political competition in state legislatures was thought to be persistent, little data had been brought together to demonstrate persistence. Persistence in relative levels of competition was further obscured by dramatic changes in absolute levels of competition. Drawing on data from a variety of historical sources, we demonstrated that political competition at the state level has been remarkably persistent over time. That is, states with above average levels of political competition in 1900–1918 had above average levels of political competition from 1866 through the 1970s. This persistence is somewhat surprising given the many changes in American society over the nineteenth and twentieth centuries. Immigration, migration, urbanization, the extension of the franchise to adults who were not white or male, and wars are just a few of the factors that might have led to political change.

Why has political competition been so persistent? Initial conditions were thought to have played a role. The Civil War had focused almost all of the attention on initial conditions related to agriculture and on the wealth of southern elites. Trade received relatively less attention. While some political historians—many of whom are cited here—had discussed tensions between rural and urban elites or different types of elites, large-scale work analyzing large numbers of states has been rare. Many other political historians have focused on voters, voting patterns, the meaning of contests, and specific histories of movements.

Our contribution has been to offer a novel hypothesis regarding the mechanism through which initial conditions shaped state legislatures and to bring together data from many different historical sources in support of this hypothesis. We hypothesized that initial conditions shaped the occupational structure of the early economic elite, and that the occupational structure of the early economic elite influenced early levels of competition in the state legislature. Economic elites influenced selection of state legislators from their geographic districts. Thus, states with more occupationally diverse economic elites had more competitive early state legislatures, because different types of elites tended to support different political parties. Both the qualitative and the quantitative historical evidence support this hypothesis. The occupational homogeneity of state elites in 1860 was persistently negatively associated with state political competition from 1866 through end of the twentieth century. This relationship remains negative and sizeable even when controls are added for other factors such as slavery.

A prominent alternative hypothesis is that wealth of the state elites influenced levels of political competition. The argument is that wealthier elites will maintain a stronger hold on political competition than less wealthy elites.[1] If this is true, political competition may be persistently negatively associated with elite wealth. In the American states, the wealth concentrations of state elites in 1860 exhibited a variable relationship with state political competition over time. When controls were added for slavery, the association between wealth of the state elites and state political competition was generally positive.

Economic and political factors drove the persistence of political institutions. Because the underlying mix of economic activity in the geographic districts within states changed slowly, the party affiliations of state legislators from those districts also changed slowly. Even as the mix of economic activities

[1] Engerman and Sokoloff (1997 and 2005) developed this argument in their study of institutions in former colonies in North and South America.

changed, it took time for new elites to supplant old elites. At the local level, old elites probably invested in maintaining the status quo for as long as possible. At the state level, the legislature and the courts tended not to change the geographic units from which legislators were elected, even when malapportionment was relatively severe. All of these factors contributed to persistence in relative levels of political competition.

Our analysis suggests that occupational homogeneity of the state elite should be recognized for the key role it played in shaping the evolution of state legislatures. Slavery and the Civil War undoubtedly played important roles in shaping both relative and absolute levels of political competition. We provide evidence that slavery did shape political competition. But our larger point is that the central role of the occupational homogeneity or heterogeneity of the state elite has largely been overlooked.

While it is impossible to know what would have happened in the absence of the Civil War, the evidence suggests that occupational homogeneity would have remained a key determinant of relative levels of political competition. Further, the same factors that led to persistence in the presence of the Civil Warwould have led to persistence in the absence of the Civil War. That is, geographic-based voting would have led to enormous persistence in political competition in the state legislature. The underlying mix of economic activities would still have changed slowly, and existing elites would still have invested in retaining political power even as their economic power waned.

Persistence and Mechanisms: State Courts

Chapters 2, 5, and 6 provided a framework within which to understand the evolution of American state courts. As with political competition, certain elements of our story have been around in the legal and historical literatures. These literatures, however, are much smaller than the literatures related to politics and political competition. The structures of state courts were known to be persistent. Drawing on data from a number of sources, we documented the extent of this persistence for a number of measures of courts. For example, states with above average levels of judicial independence in 1900–1918 had above average levels of judicial independence from 1866 to 2000. The expenditure data for courts covers a shorter span, 1952–2000, but shows considerable persistence as well. Further, researchers have constructed models to describe the conditions under which state legislatures might make changes to levels of judicial independence.

Why has court structure been so persistent? One reason for persistence is that changing the retention methods and tenure of high court judges requires changing the state constitution. This raises the cost of change, but the many changes made to state constitutions suggest that this cost is not prohibitive. Other factors must have also played a role. Initial conditions were thought to have played a role, particularly the timing of entry, since states that entered the union before 1847 all entered with appointment systems. Other factors such as slavery may have been recognized as well. The potential role of colonial legal systems had either been overlooked or dismissed.

Our contribution has been to offer a novel hypothesis regarding the mechanism through which colonial legal systems shaped state courts and to bring together data from many different historical sources in support of this hypothesis. We hypothesized that colonial legal systems, acting through their influence on individuals, shaped the early balance of power between the legislature and the judiciary. Certainly by the French Revolution (1789) and possibly earlier, the civil-law legal system conceived of the legislature as the dominant branch of government. The judiciary was relatively subordinate in the sense that its role was to enforce laws written by the legislature. In the common-law legal system by that time, the legislature and the courts were relatively more equal. Recall that twelve of the thirteen states classified as civil law adopted common law around the time of statehood.

If colonial legal systems had an influence on the balance of power between the legislature and the courts, the balance would favor the legislature more in civil-law states and would be more equal in common-law states. Historical evidence shows that individuals with civil-law backgrounds or affiliations were active in politics around the time the balance of power was struck. Although individuals with civil-law backgrounds probably played a formative role, individuals with common-law backgrounds might also have preferred a more subordinate judiciary.

We presented a model that linked political competition in the state legislature and colonial legal system to levels of judicial independence. One implication of the model is that the threshold level of political competition at which civil-law and common-law states will change judicial independence differs. Thus, conditional on the level of political competition, average levels of judicial independence will differ in civil-law and common-law states. Empirically, the threshold levels of political competition associated with increases in judicial independence were higher in civil-law states than in common-law states. Controlling for levels of political competition, civil law was persistently negatively associated with judicial independence from 1866 through 2000. This relationship remained negative and sizeable even when controls were added for other factors such as slavery and entry into the union before 1847.

Civil-law and common-law states differed in other dimensions that are consistent with their legal origins. Civil-law states adopted intermediate appellate courts more quickly, seemingly to better monitor lower courts, and funded their state courts less generously than common-law states did. Civil-law and common-law states differed in how they adjusted expenditures on state courts following changes in judicial retention. Common-law states increased expenditures following changes in judicial retention. In civil-law states the increases tended to be smaller or even zero. The differences in their responses suggest differences in their views of judicial reform. If civil-law legislatures believed that reform gave judges "too much independence," then lower budgets would be a way to rein in judicial behavior at the margin.

The model highlights some of the reasons for persistence in levels of judicial independence. For levels of political competition below the threshold, pc^*, legislatures maintain low levels of independence, and for levels of competition above the threshold, pc^*, legislatures maintain high levels of independence. If political competition only rarely rises above or falls below pc^*, then changes in the level of judicial independence will be rare. This dynamic, combined with the cost of changing the state constitution, meant that states rarely changed judicial retention or tenure more than one or twice over the period 1866–2000.

Politics, Courts, and Economic Outcomes

In chapter 1, figure 1.1 outlined the relationships among political competition in the state legislature, judicial independence in state high courts, and state per capita income, which is a measure of long-run growth. This book argues that the relationships between initial conditions and institutions were causal.

A challenge that arises in arguing institutions caused state per capita income is that many of the same initial conditions that shaped political competition are likely to have influenced state per capita income. Without a better understanding of the specific mechanisms through which initial conditions and political institutions acted on income, it is impossible to disentangle the two to make a causal statement about the effects of political institutions.

The first two columns of table 7.1 show the relationships between institutions and state per capita income in 2003 with and without controls for slavery. The results need to be interpreted cautiously. The number of observations is

TABLE 7.1.
POLITICAL AND LEGAL INSTITUTIONS AND PER CAPITA INCOME IN 2003

Dependent variable	*Log PCI 2003*	*Log PCI 2003*	*Log PCI 2003*	*Log PCI 2003*
Log of PCI, 1900	0.131**	0.190**	0.223***	0.366***
	(0.0572)	(0.0730)	(0.0507)	(0.0719)
Log of slavery, 1860		0.0312*		0.0451***
		(0.0169)		(0.0160)
Ranney, 1900–2000	0.0002	0.0009		
	(0.0010)	(0.0011)		
Log of occupational homogeneity, 1860			−0.0113	−0.0088
			(0.0607)	(0.0541)
Judicial independence, 1900–2000	0.0206***	0.0213***		
	(0.00688)	(0.00665)		
Civil			−0.0967***	−0.109***
			(0.0286)	(0.0320)
Constant	9.105***	8.551***	8.545***	7.311***
	(0.421)	(0.584)	(0.619)	(0.753)
Observations	46	46	27	27
R^2	0.406	0.445	0.673	0.734

Notes: Standard errors are heteroskedasticity robust. The notations *, **, and *** denote significance at the 10 percent, 5 percent, and 1 percent levels. The base year for deflating income is 2000. Nebraska is dropped because it has a unicameral legislature for most of the twentieth century and Louisiana is dropped because it is has maintained a civil law system after entering the union. Judicial independence, 1900–2000 is the average level of the judicial independence (scaled 1–9) developed by Epstein, Knight, and Shvetsova (2002).

small, and the measures of both political competition in the state legislature and independence of judges on the state high courts at best crudely capture the nature of these complex institutions. Further, per capita income in 1900 may not completely control for the effects of agriculture and trade on income in 2003.

States with higher per capita income in 1900, higher shares of slaves in 1860, and greater judicial independence from 1900 to 2000 had higher per capita income in 2003. The association between the share of slaves and income captures the catching up of southern states during the twentieth century, particularly after World War II. Measures of state political competition were not independently important for state per capita income in 2003. Greater judicial independence over the period 1900–2000 was associated with higher per capita income in 2003. More independent judges may contribute to higher income by reigning in the behavior of the state legislature and other special interests and by making decisions that expand economic activity in aggregate. Indeed, independent judges may do what Massachusetts residents desired in the seventeenth century: "Control the discretion of the governing elite and thereby prevent it from interfering unpredictably in the lives of ordinary people."[2]

The next two columns of table 7.1 replace the measures of political and legal institutions with historical measures—the occupational homogeneity of the elite in 1860 and civil law. The patterns are very similar to the patterns in the first two columns. States with higher per capita income in 1900 and higher shares of slaves in 1860 had higher per capita income in 2003. Civil-law states had lower per capita income, and occupational homogeneity of the elite was not significant.

In sum, judicial independence was positively associated with state per capita income in 2003. The fact that similar factors influenced state political institutions and state per capita income in 1900 made it difficult to disentangle their effects.

Future Research

A number of topics raised in this book warrant more research. It is worth mentioning two, outcomes and reform. We have had little to say about how political competition affects policy outcomes and about how court structure affects legal outcomes, yet these outcomes are clearly important. Although other scholars

[2] Nelson (2008), 78.

have investigated aspects of these issues, more remains to be done. Our work has some possible implications for states or countries interested in reforming their legal institutions. Here too, more work remains to be done.

Outcomes

The relationship between political competition and policy outcomes has been of interest to some scholars for decades. What we have in mind is more detailed analysis of more outcomes over longer periods of time than has previously been done to try to determine the conditions under which political competition leads to "good" policy outcomes. For example, education as embodied in human capital is a critical input into economic activity, and state legislatures had a large influence on education policy. One could study the relationship between state political competition and state educational outcomes as measured by passage of legislation, state funding, and school attendance from 1866–2000 and possibly for earlier periods. In principle such a study is possible. Schooling and labor laws are fairly well cataloged, but one would want to look at other education-related laws. Annual data on state funding would need to be compiled for some periods, particularly in the nineteenth century. Annual school attendance numbers are available, but would need to be digitized. One could imagine doing similar studies that examined the relationship between political competition and policy outcomes in other key areas such as infrastructure investment, public health, and banking.

A series of studies such as these would inform discussions of when political competition does or does not lead to beneficial outcomes. They could also shed light on a related issue, which is whether the party in power also had an influence on outcomes. Suppose two hypothetical states had highly competitive legislatures, but one was controlled by the Democrats, while the other was controlled by the Republicans. Was one party more likely to invest in education or infrastructure or public health or a strong banking system? And if so, which party was investing in which categories and during what time periods? All of this could lead to a more nuanced understanding of politics, policies, and growth.

Legal scholars have also long been interested in the relationship between court structure and legal outcomes. One of the key difficulties is measuring legal outcomes. A variety of key statistics are available on the composition of filings and dispositions and on issues such as delay and backlog. Unfortunately, even these types of statistics tend to be available only for the late twentieth century. It is difficult to make inferences about the quality of legal outcomes using these statistics. Extensive delay is probably not a sign of a well-functioning legal system. Were the decisions unbiased? How closely did they adhere to the current understanding of case law? Because many possible cases are never filed and because many cases that are filed never come to trial, it is also difficult to know how cases that reach trial are related to cases that do not reach trial. Detailed analysis of how state courts handle specific types of cases is possible, but extraordinarily time consuming. Because of this, scholars have relied on cross-court citations as a metric of court quality, sometimes in conjunction with other measures such as the number and length of opinions and patterns of judicial voting.

A more detailed analysis of long-run relationships between court structure and outcomes requires the development of measures of outcomes that can be (retrospectively) computed for large numbers of courts over long periods of time and are plausibly related to court quality. In the absence of that, existing or new survey-based measures can offer insight. They will, however, need to be conducted frequently and over a relatively large time horizon before they can generate insights into how changes in court structure lead to changes in outcomes.

Reform

The analysis in this chapter and in some previous chapters hints at possible lessons for reform. For example, states or countries should increase the independence of their judiciary by doing away with competitive retention elections, establishing intermediate appellate courts if they do not already have them, extending term lengths, and possibly increasing the funding of courts. If opportunities present themselves, governments should take measures that will facilitate greater political competition.

One wants to be cautious in drawing lessons for reform, however, because most of what is known about the quality of political and legal institutions and per capita income comes from cross-sectional analysis. Time series analysis would be more convincing. Ideally, changes in political and legal institutions would occur for exogenous reasons that would lead to observable growth effects, allowing causal inferences to be made.

To the extent that we have something concrete to say about reform, it is that policymakers interested in change would benefit from understanding history. Political and legal institutions are extremely persistent for reasons that have been highlighted in the book. This is not to say that reform is impossible. The average level of state political competition has risen in the last fifty years, driven by a variety of societal and other factors. State court systems have evolved as well, as legislatures changed judicial retention and term length, adopted intermediate appellate courts, and adjusted funding. Some changes have been hotly contested, and some proposed changes have failed to be adopted. A better understanding of how and why institutions are persistent can inform efforts at change.

References

Abbott, John Stevens Cabot. 1875. *The History of the State of Ohio: From the Discovery of the Great Valley, to the Present Time*. Detroit: Northwestern Publishing Company.

———. 1892. *The History of Maine*. 2nd ed. Revised throughout and five chapters of new material added by Edward H. Elwell. Portland, Me: Published for E.E. Knowles & Co. by B. Thurston Co.

Acemoglu, Daron, Simon Johnson, and James A. Robinson. 2001. "The Colonial Origins of Comparative Development: An Empirical Investigation." *American Economic Review* 91: 1369–1401.

Acemoglu, Daron, and James A. Robinson. 2006. "Economic Backwardness in Political Perspective." *American Political Science Review* 100: 115–131.

———. 2008. "Persistence of Power, Elites, and Institutions." *American Economic Review* 98: 267–293.

Alvord, Clarence Walworth. 1909. *Kaskaskia Records, 1778–1790*. Springfield, Ill.: Trustees of the Illinois State Historical Library.

———. 1920. *The Illinois Country 1673–1818*. Springfield, Ill.: Illinois Centennial Commission.

Andrews, Israel De Wolf. 1853. *Communication from the secretary of the Treasury, transmitting, in compliance with a resolution of the Senate of March 8, 1851, the report of Israel D. Andrews . . . on the trade and commerce of the British North American colonies, and upon the trade of the Great lakes and rivers; also, notices of the internal improvements in each state, of the gulf of Mexico and straits of Florida, and a paper on the cotton crop of the United States*. Washington, D.C.: Robert Armstrong.

Ansolabehere, Stephen, and James M. Snyder. 2002. "The Incumbency Advantage in U.S. Elections: An Analysis of State and Federal Offices, 1942–2000." *Election Law Journal* 1: 313–38.

Archives of Spanish Government of West Florida, 1782–1816. National Archives T1116.

Armstrong, Edward, ed. 1969. *Record of Upland Court from the 14th of November, 1676, [to] the 14th of June, 1681*. Salt Lake City: Filmed by the Genealogical Society of Utah.

Arnold, Anna Estelle. 1914. *A History of Kansas*. Topeka: State of Kansas.

Arnold, Morris S. 1985. *Unequal Laws Unto a Savage Race: European Legal Traditions in Arkansas, 1686–1836*. Fayetteville: University of Arkansas Press.

Arthur, Timothy Shay, and William Henry Carpenter. 1852. *The History of Kentucky: From Its Earliest Settlement to the Present Time*. Philadelphia: Lippincott, Grambo.

———. 1853. *The History of Georgia: From Its Earliest Settlement to the Present Time*. Philadelphia: Lippincott, Grambo.

Ashe, Samuel A'Court. 1908. *History of North Carolina*. Greensboro, N.C.: C.L. Van Noppen.

Ashmore, Harry S. 1984. *Arkansas: A History*. New York: Norton.

Atack, Jeremy, and Fred Bateman. 1981. "Egalitarianism, Inequality, and Age: The Rural North in 1860." *Journal of Economic History* 41: 85–93.

Atack, Jeremy, Fred Bateman, Michael Haines, and Robert Margo. 2009. "Did Railroads Induce or Follow Economic Growth? Urbanization and Population Growth in the American Midwest, 1850–60." NBER Working Paper No. 14640.

Atwater, Caleb. 1838. *A History of the State of Ohio: Natural and Civil*. Cincinnati: Glezen & Shepard.

Austin, George Lowell. 1875. *The History of Massachusetts: From the Landing of the Pilgrims to the Present Time. Including a Narrative of the Persecutions by State and Church in England; the Early Voyages to North America; the Explorations of the Early Settlers; Their Hardships, Sufferings and Conflicts with the Savages*. Boston: B.B. Russell.

Baker, J. H. 2007. *An Introduction to English Legal History*. 4th ed. Oxford: Oxford University Press.

Bakken, Gordon Morris. 1974. "Contract Law in the Rockies, 1850–1912." *American Journal of Legal History* 18: 33–51.

———, ed. 2000. *Law in the Western United States*. Norman: University of Oklahoma Press.

Bancroft, Hubert Howe. 1888. *History of California*. Reprint, Santa Barbara: Wallace Hebberd, 1970.

Bancroft, Hubert Howe, and Alfred Bates. 1889. *History of Utah: 1540–1886*. San Francisco: History Company.

Bancroft, Hubert Howe, and Henry Lebbeus Oak. 1886. *History of the Northwest Coast*. San Francisco: History Company.

———. 1889. *History of Arizona and New Mexico: 1530–1888*. San Francisco: History Company.

Bancroft, Hubert Howe, and Frances Fuller Victor. 1890a. *History of Nevada, Colorado, and Wyoming, 1540–1888*. San Francisco: History Company.

———. 1890b. *History of Washington, Idaho and Montana*. San Francisco: History Company.

Banerjee, Abhijit, and Lakshmi Iyer. 2005. "History, Institutions and Economic Performance: The Legacy of Colonial Land Tenure Systems in India." *American Economic Review* 95: 1190–1213.

Banner, Stuart. 2000. *Legal Systems in Conflict: Property and Sovereignty in Missouri, 1750–1860*. Norman: University of Oklahoma Press.

Barbé-Marbois, François. 1830. *The History of Louisiana: Particularly of the Cession of that Colony to the United States of America: with an Introductory Essay on the*

Constitution and Government of the United States. Trans. William Beach Lawrence. Philadelphia: Carey & Lea.

Barstow, George. 1842. *The History of New Hampshire: From Its Discovery, in 1614, to the Passage of the Toleration Act, in 1819*. Concord, N.H.: I.S. Boyd.

Baum, Dale D. 1984. *The Civil War Party System: The Case of Massachusetts, 1848–1876*. Chapel Hill: University of North Carolina Press.

Beard, Charles Austin. 1935. *An Economic Interpretation of the Constitution of the United States*. New York: Macmillan.

Becker, Daniel, and Malia Reddick. 2003. *Judicial Selection Reform: Examples from Six States*. Des Moines, Iowa: American Judicature Society.

Belknap, Jeremy, and John Farmer. 1862. *The History of New-Hampshire*. Dover, N.H.: G. Wadleigh.

Bell, John. 1988. "Principles and Methods of Judicial Selection in France." *Southern California Law Review* 61:1757–1794.

Benson, Lee. 1955. *Merchants, Farmers, and Railroads: Railroad Regulation and New York Politics, 1850–1887*. Cambridge: Harvard University Press.

———. 1960. *Turner and Beard: American Historical Writing Reconsidered*. Glencoe, Ill.: Free Press.

———. 1961. *The Concept of Jacksonian Democracy: New York as a Test Case*. Princeton: Princeton University Press.

Berkowitz, Daniel, and Karen Clay. 2005. "American Civil Law Origins: Implications for State Constitutions and State Courts." *American Law and Economics Review* 7: 62–84.

———. 2006. "The Effect of Judicial Independence on Courts: Evidence from the American States." *Journal of Legal Studies* 35: 399–400.

Berkowitz, Daniel, and David N. DeJong. 2011. "Growth in Post-Soviet Russia: A Tale of Two Transitions." Forthcoming in the *Journal of Economic Behavior and Organization*.

Berman, David. 1988. "Political Culture, Issues, and the Electorate: Evidence from the Progressive Era." *Western Political Quarterly* 41: 169–180.

Berman, Harold J. 1983. *Law and Revolution: The Formation of the Western Legal Tradition*. Cambridge: Harvard University Press.

Besley, Timothy, and Abigail Payne. 2003. "Judicial Accountability and Economic Policy Outcomes: Evidence from Employment Discrimination Charges." IFS Working Papers W03/11, Institute for Fiscal Studies, revised.

Besley, Timothy, Torsten Persson, and Daniel Sturm. 2010. "Political Competition, Policy and Growth: Theory and Evidence from the United States." *Review of Economic Studies* 77: 1329–1352.

Bettersworth, John Knox. 1959. *Mississippi: A History*. Austin, Tex.: Steck.

Beverley, Robert, and Charles Campbell. 1722. *The History of Virginia: In Four Parts*. 2nd ed. Reprint, Richmond: J. W. Randolph, 1855.

Bohn, H., and R. Inman. 1996. "Balanced Budget Rules and Public Deficits: Evidence from the U.S. States." NBER Working Paper No. 5533.

Bonneau, Chris W. 2005. "Electoral Verdicts: Incumbent Defeats in State Supreme Court Elections." *American Politics Research* 33: 818–841.

The Book of the States. Various years. Lexington, Ky.: Council of State Government.

Bourke, Paul, and Donald A. DeBats. 1995. *Washington County: Politics and Community in Antebellum America*. Baltimore: Johns Hopkins University Press.

Bozman, John Leeds. 1837. *The History of Maryland: From Its First Settlement, in 1633, to the Restoration, in 1660*. Baltimore: J. Lucas & E. K. Deaver.

Brace, Paul, and Melinda Gann Hall. 1997. "The Interplay of Preferences, Case Facts, Context, and Rules in the Politics of Judicial Choice." *Journal of Politics* 59: 1206–1231.

Bradford, Alden. 1835. *History of Massachusetts, for Two Hundred Years: From the Year 1620 to 1820*. Boston: Hilliard, Gray.

Brewer, Willis. 1872. *Alabama, Her History, Resources, War Record, and Public Men: From 1540 to 1872*. Montgomery: Barrett & Brown.

Briggs, Winstanley. 1990. "Le Pays Des Illinois." *William and Mary Quarterly* 47: 30–56.

Brodhead, John R., and E. B. O'Callaghan. 1853–1887. *Documents Relative to the Colonial History of the State of New York: Procured in Holland, England, and France*. Vol. 12, *Documents Relating to the History of the Dutch and Swedish Settlements on the Delaware River*. Albany, N.Y. Weed, Parsons.

Browne, J. Ross. 1850. *Report of the Debates in the Convention of California, on the Formation of the State Constitution, In September and October, 1849*. Reprint, New York: Arno Press, 1973.

Burbank, Stephen B., and Barry Friedman, eds. 2002. *Judicial Independence at the Crossroads: An Interdisciplinary Approach*. Thousand Oaks, Calif.: Sage.

Burnham, W. Dean, Jerome M. Clubb, and William Flanigan. 1991. *State-Level Congressional, Gubernatorial and Senatorial Election Data for the United States, 1824–1972 [Computer File]. ICPSR ed.* Ann Arbor, MI: Inter-university Consortium for Political and Social Research.

Burns, Robert Ignatius, ed. 2001. *Las Siete Partidas*. Vol. 1, *The Medieval Church: The World of Clerics and Laymen*. Trans. Samuel Parsons Scott. Philadelphia: University of Pennsylvania Press.

Burton, Clarence Monroe, William Stocking, and Gordon K. Miller. 1922. *The City of Detroit, Michigan, 1701–1922*. Detroit: S.J. Clarke.

Butte, George C. 1917. "Early Development of Law and Equity in Texas." *Yale Law Journal* 26: 699–709.

Callahan, James Morton. 1913. *Semi-centennial History of West Virginia: With Special Articles on Development and Resources*. Charleston: Semi-centennial Commission of West Virginia.

Campbell, Ballard C. 1980. *Representative Democracy: Public Policy and Midwestern Legislatures in the Late Nineteenth Century*. Cambridge: Harvard University Press.

Canon, Bradley C. 1972. "The Impact of Formal Selection Processes on the Characteristics of Judges—Reconsidered." *Law and Society Review* 6: 579–593.

Cardman, Denise. 2007. "Select ABA Policies Relating to State Judicial Independence." Available at http://www.abanet.org/nabe/govrelations/workshop/judindpolicies.pdf.

Carey, Charles Henry. 1922. *History of Oregon*. Portland: Pioneer Historical Publishing Company.

Carey, W. P. 2009. "Sandra Day O'Connor: Where Judges Can Be Bought and Sold." January 28. Available at http://knowledge.wpcarey.asu.edu/article.cfm?articleid=1739#.

Chapp, Joy A., and Roger A. Hanson. 1990. "Intermediate Appellate Courts: Improving Case Processing: Final Report / Submitted to the State Justice Institute by National Center for State Courts. April. Williamsburg, Va.: The Center.

Choi, Stephen J., G. Mitu Gulati, and Eric A. Posner. 2008. "Professionals or Politicians: The Uncertain Empirical Case for an Elected Rather Than Appointed Judiciary." *Journal of Law, Economics, and Organizations*. Advance Access published on November 5, DOI 10.1093/jleo/ewn023.

Claiborne, John Francis Hamtramck. 1880. *Mississippi, as a Province, Territory, and State: With Biographical Notices of Eminent Citizens*. Jackson: Power & Barksdale.

Clark, George Larkin. 1914. *A History of Connecticut: Its People and Institutions*. New York: G.P. Putnam's Sons.

Clark, Thomas Dionysius, and John D. W. Guice. 1996. *The Old Southwest, 1795–1830*. Norman: University of Oklahoma Press.

Coatsworth, John H. 1993. "Notes on the Comparative Economic History of Latin America and the United States." In Walter L. Bernecker and Hans Werner Tobler, ed., *Development and Underdevelopment in America: Contrasts in Economic Growth in North and Latin America in Historical Perspective*. New York: Walter de Gruyter.

Coker, William. 1999. "Pensacola 1686–1821." In Judith Ann Bense, ed., *Archaeology of Colonia Pensacola*. Gainesville: University Press of Florida.

Cole, Cyrenus. 1921. *A History of the People of Iowa*. Cedar Rapids: Torch Press.

Collins, Edward Day. 1903. *A History of Vermont, with Geological and Geographical Notes, Bibliography, Chronology*. Boston: Ginn.

Colorado State Archives. 2001. "Spanish-Mexican Land Grants: A Brief Introduction." Available at http://www.colorado.gov/dpa/doit/archives/mlg/mlg.html.

Congressional, Gubernatorial and Senatorial Election Data for the United States, 1824–1972 [Computer File]. ICPSR ed. Ann Arbor, MI: Inter-university Consortium for Political and Social Research.

Cooper, Leigh G. 1920. "Influence of the French Inhabitants of Detroit on its Early Political Life." *Michigan History Magazine* 4: 299–304.

Coutant, Charles Griffin. 1899. *The History of Wyoming from the Earliest Known Discoveries: In Three Volumes*. Laramie: Chaplin, Spafford & Mathison.

Crosby, Elisha O. 1945. *Memoirs of Elisha Oscar Crosby: Reminiscences of California and Guatemala from 1849 to 1864*. San Marino, Calif.: Huntington Library.

Cummins, Light Townsend, Bennett H. Wall, and Judith Kelleher Schafer. 2002. *Louisiana: A History*. 4th ed. Wheeling, Ill.: Harlan Davidson.

Cutter, Charles R. 1995. *The Legal Culture of Northern New Spain, 1700–1810*. Albuquerque: University of New Mexico Press.

Dalzell, Robert F., Jr. 1987. *Enterprising Elite: The Boston Associates and the World They Made*. Cambridge: Harvard University Press.

Dargo, George. 1975. *Jefferson's Louisiana: Politics and the Clash of Legal Traditions*. Cambridge: Harvard University Press.

Davis, Lance, and Douglass C. North. 1971. *Institutional Change and American Economic Growth*. New York: Cambridge University Press.

Davis, Samuel Post, ed. 1913. *The History of Nevada*. Reno: Elms Publishing.

Dawson, J. 1960. *A History of Lay Judges*. Cambridge: Harvard University Press.

Day, Christine L., and Charles D. Hadley. 2001. "Feminist Diversity: The Policy Preferences of Women's PAC Contributors." *Political Research Quarterly* 54: 673–686.

Dealy, Glen. 1968. "Prolegomena on the Spanish American Political Tradition." *Hispanic American Historical Review* 48: 37–58.

Dell, Melissa. 2009. "The Persistent Effects of Peru's Mining *Mita*." *Econometrica* 78: 1863–1903.

Derden, James. 1910. *Digest of the Decisions of the Supreme Court of New Mexico: Volumes 1 to 14, Inclusive: and All New Mexico Decisions in Pacific Reporter, 1 to 106, Inclusive: with Table of Cases and with Rules of the Supreme and District Courts*. Denver: W.H. Courtright.

Diamond, Jared. 1997. *Guns, Germs and Steel: The Fates of Human Societies*. New York: Norton.

Dillon, John Brown. 1859. *A History of Indiana from Its Earliest Exploration by Europeans to the Close of the Territorial Government in 1816: Comprehending a History of the Discovery, Settlement, and Civil and Military Affairs of the Territory of the U.S. Northwest of the River Ohio, and a General View of the Progress of . . .* Indianapolis: Bingham & Doughty.

Doerflinger, Thomas M. 1986. *A Vigorous Spirit of Enterprise: Merchants and Economic Development in Revolutionary Philadelphia*. Williamsburg: Institute of Early American History and Culture.

Douglas, James W., and Roger E. Hartley. 2003. "The Politics of Court Budgeting in the States: Is Judicial Independence Threatened by the Budgetary Process?" *Public Administration Review* 63: 441–454.

Dubois, Philip. 1980. *From Ballot to Bench: Judicial Elections and the Quest for Accountability*. Austin: University of Texas Press.

Dunbar, Willis Frederick, and George S. May. 1995. *Michigan*. 3rd ed. Grand Rapids: Eerdmans.

Dwight, Theodore. 1842. *The History of Connecticut: From the First Settlement to the Present Time*. New York: Harper & Brothers.

Easterly, William. 2006. *The White Man's Burden: The Wacky Ambition of the West to Transform the Rest*. New York: Penguin.

Easterly, William, and Ross Levine. 2003. "Tropics, Germs, and Crops: The Role of Endowments in Economic Development." *Journal of Monetary Economics* 50: 3–39.

Ekberg, Carl J. 1998. *French Roots in the Illinois Country: The Mississippi Frontier in Colonial Times*. Urbana: University of Illinois Press.

Elazar, Daniel. 1984. *American Federalism: A View from the States*. 2nd ed. New York: HarperCollins.

Engerman, Stanley L., Elisa Mariscal, and Kenneth L. Sokoloff. 1998. "Schooling, Suffrage and the Persistence of Inequality in the Americas, 1800–1945." Working paper, Department of Economics, UCLA.

Engerman, Stanley L., and Kenneth L. Sokoloff. 1997. "Factor Endowments, Institutions and Differential Paths of Growth among the New World Economies." In Stephen Haber, ed., *How Latin America Fell Behind*. Stanford: Stanford University Press.

———. 2000. "History Lessons: Institutions, Factors Endowments, and Paths of Development in the New World." *Journal of Economic Perspectives* 14: 217–232.

———. 2005. "Colonialism, Inequality, and Long-Run Paths of Development." NBER Working Paper No. 11057, January.

Epstein, Lee, Jack Knight, and Olga Shevtsova. 2002. "Selecting Selections Systems." In S. B. Burbank and Barry Friedman, eds., *Judicial Independence at the Crossroads: An Interdisciplinary Approach*. Thousand Oaks, Calif.: Sage.

Esarey, Logan. 1915. *A History of Indiana*. Indianapolis: W.K. Stewart.

Fairbanks, George Rainsford. 1901. *Florida, Its History and Its Romance: The Oldest Settlement in the United States, Associated with the Most Romantic Events of American History, Under the Spanish, French, and American Flags, 1497–1901*. Jacksonville: H. and W. B. Drew.

Farish, Thomas Edwin. 1916. *History of Arizona*. San Francisco: Filmer Brothers Electrotype Company Typographers and Stereotypers.

Fehrenbach, T. R. 2000. *Lone Star: A History of Texas and the Texans*. Cambridge, Mass.: Da Capo Press.

Fernandez, Mark F. 2001. *From Chaos to Continuity: The Evolution of Louisiana's Judicial System, 1712–1862*. Baton Rouge: Louisiana State University Press.

Fernow, Berthold, ed. 1897. *The Records of New Amsterdam from 1653 to 1674 Anno Domini*. Trans. Edmund Bailey O'Callaghan. 7 vols. Reprint, Baltimore: Genealogical Publishing, 1976.

Fisher, Louis. 1998. *Constitutional Dialogues: Interpretation as Political Process*. Princeton: Princeton University Press.

Fishlow, Albert. 1965. *American Railroads and the Transformation of the Ante-Bellum Economy*. Cambridge: Harvard University Press.

Fogel, Robert W. 1964. *Railroads and American Economic Growth: Essays in Econometric History*. Baltimore: Johns Hopkins Press.

Folwell, William Watts. 1921. *A History of Minnesota*. Saint Paul: Minnesota Historical Society.

Ford, Lacy K. 1984. "Rednecks and Merchants: Economic Development and Social Tensions in the South Carolina Upcountry, 1865–1900." *Journal of American History* 71: 294–318.

Formisano, Ronald P. 1983. *The Transformation of Political Culture: Massachusetts Parties, 1790s–1840s*. New York: Oxford University Press.

———. 1999. "The 'Party Period' Revisited." *Journal of American History* 86: 93–120.

———. 2001. "The Concept of Political Culture." *Journal of Interdisciplinary History* 31: 393–426.

Friedman, Lawrence M. 1986. *A History of American Law*. 2nd ed. New York: Simon and Schuster.

———. 1988. "State Constitutions in Historical Perspective." *Annals of the American Academy of Political Science* 496: 33–42.

Gallup, John Luke, Jeffrey D. Sachs, and Andrew D. Mellinger. 1998. "Geography and Economic Development." In Boris Pleskovic and Joseph E. Stiglitz, eds., *Annual World Bank Conference on Development Economics*. Washington, D.C.: World Bank.

Gardner, James A. 2006. "Representation Without Party: Lessons from State Constitutional Attempts to Control Gerrymandering." *Rutgers Law Journal* 37: 881–970.

Gaston, Joseph, and George H. Himes. 1912. *The Centennial History of Oregon, 1811–1912*. Chicago: S. J. Clarke.

Gehring, Charles T. 1981. *Delaware Papers*. Vol. 2. Baltimore: Genealogical Publishing.

General Land Office. 2009. "History of Texas Public Lands." Available at http://www.glo.state.tx.us/archives/history/republic_texas.html.

Gibson, Arrell Morgan, and Victor Emmanuel Harlow. 1984. *The History of Oklahoma*. Norman: University of Oklahoma Press.

Glaeser, Edward, and Janet Kohlhase. 2003. "Cities, Regions, and the Decline of Transport Costs." NBER Working Paper No. 9886.

Glaeser, Edward, and Andrei Shleifer. 2002. "Legal Origins." *Quarterly Journal of Economics* 117: 1932–1230.

Glick, Henry R., and Craig Emmert. 1987. "Selection Systems and Judicial Characteristics: The Recruitment of State Supreme Court Justices." *Judicature* 70: 228–35.

Goldberg, Ellis, Eric Wibbels, and Eric Myukiyehe. 2008. "Lessons from Strange Cases: Democracy, Development, and the Resource Curse in the U.S. States." *Comparative Political Studies* 41: 477–514.

Gómez, Laura E. 2007. *Manifest Destinies: The Making of the Mexican American Race*. New York: NYU Press.

Goodman, Paul. 1986. "The Social Basis of New England Politics in Jacksonian America." *Journal of the Early Republic* 6: 23–58.

Gordon, Thomas Francis. 1829. *The History of Pennsylvania: From Its Discovery by Europeans, to the Declaration of Independence in 1776*. Philadelphia: Carey, Lea & Carey.

———. 1834. *The History of New Jersey: From Its Discovery by Europeans, to the Adoption of the Federal Constitution*. Trenton: D. Fenton.

Greene, George Washington. 1877. *A Short History of Rhode Island*. Providence: J. A. & R. A. Reid.

Greene, Jack P., and J. R. Pole. 2003. *A Companion to the American Revolution*. Malden, Mass.: Blackwell.

Greif, Avner. 2006. *Institutions and the Path to the Modern Economy: Lessons from Medieval Trade*. New York: Cambridge University Press.

Gryzmala-Busse, Anna. 2007. *Rebuilding Leviathan: Party Competition and State-Exploitation in Post-Communist Democracies*. New York: Cambridge University Press.

Haber, Stephen, and Victor Menaldo. 2009. "Do Natural Resources Fuel Authoritarianism? A Reappraisal of the Resource Curse." Working paper.

Hailey, John. 1910. *The History of Idaho*. Boise: Press of Syms-York Co.

Hall, Hiland. 1868. *The History of Vermont, from Its Discovery to Its Admission Into the Union in 1791*. Albany, N.Y.: J. Munsell.

Hall, Kermit. 1984. "The Impact of Popular Election on the Southern Appellate Judiciary." In David J. Bodenhamer and James W. Ely, Jr., eds., *Ambivalent Legacy: A Legal History of the South*. Jackson: University Press of Mississippi.

———. 1987. "Dissent on the California Supreme Court, 1850–1920." *Social Science History* 11: 63–83.

———. 2006. "Judicial Independence and the Majoritarian Difficulty." In Kermit Hall and Kevin McGuire, eds. *Institutions of American Democracy: The Judicial Branch*. Oxford: Oxford University Press.

Hall, Melinda Gann. 1987. "Constituent Influence in State Supreme Courts: Conceptual Notes and a Case Study." *Journal of Politics* 49: 1117–1124.

———. 1992. "Electoral Politics and Strategic Voting in State Supreme Courts." *Journal of Politics* 54: 427–446.

———. 1995. "Justices as Representatives: Elections and Judicial Politics in the American States." *American Politics Quarterly* 23: 485–503.

———. 2001. "State Supreme Courts in American Democracy: Probing the Myths of Judicial Reform." *American Political Science Review* 95: 315–330.

Hamilton, Peter J. 1910. *Colonial Mobile: A Study of Southwestern History*. Boston: Houghton Mifflin.

Hammons, Christopher W. 1999. "Was James Madison Wrong? Rethinking the American Preference for Short, Framework-Oriented Constitutions." *American Political Science Review* 93: 837–49.

Hanson, Roger. 2001. "When Were the Intermediate Courts of Appeal Established?" National Center for State Courts. http://contentdm.ncsconline.org/cgi-bin/showfile.exe?CISOROOT=/appellate&CISOPTR=176.

Hanssen, F. Andrew. 1999. "The Effect of Judicial Institutions on Uncertainty and the Rate of Litigation: The Election versus Appointment of State Judges." *Journal of Legal Studies* 28: 205–232.

———. 2002. "On the Politics of Judicial Selection: Lawyers and State Campaigns for the Merit Plan." *Public Choice* 110: 79–97.

———. 2004a. "Learning About Judicial Independence: Institutional Change in the State Courts." *Journal of Legal Studies* 33: 431–473.

———. 2004b. "Is There a Politically Optimal Level of Judicial Independence?" *American Economic Review* 94: 712–729.

Hardy, Leroy, Alan Heslop, and Stuart Anderson. 1981. *Reapportionment Politics: The History of Redistricting in the 50 States*. Beverly Hills: Sage.

Harvey, Lashley G. 1949. "Some Problems of Representation in State Legislatures." *Political Research Quarterly* 2: 265–271.

Haywood, John, Arthur St. Clair, and Zella Armstrong Colyar. 1891. *The Civil and Political History of the State of Tennessee from Its Earliest Settlement Up to the Year 1796: Including the Boundaries of the State*. Nashville: W.H. Haywood.

Herndon, Dallas Tabor. 1922. *Centennial History of Arkansas*. Little Rock: S. J. Clarke.

Hoffer, Peter Charles. 1992. *Law and People in Colonial America*. Baltimore: Johns Hopkins University Press.

Holbrook, Thomas M., and Emily van Dunk. 1993. "Electoral Competition in the American States." *American Political Science Review* 87: 955–962.

Holloway, John N. 1868. *History of Kansas: From the First Exploration of the Mississippi Valley, to Its Admission Into the Union: Embracing a Concise Sketch of Louisiana; American Slavery, and Its Onward March; the Conflict of Free and Slave Labor in the Settlement of Kansas, and the Overthrow of the Latter, with All . . .* Lafayette, Ind.: James, Emmons.

Holmes, Jack D. L. 1963. "Law and Order in Spanish Natchez, 1781–1798." *Journal of Mississippi History* 25: 186–201.

Holt, Michael F. 1983. *The Political Crisis of the 1850s*. New York: Norton.

Horwitz, Morton J. 1977. *The Transformation of American Law, 1780–1860*. Cambridge: Harvard University Press.

Houck, Louis. 1908. *A History of Missouri from the Earliest Explorations and Settlements Until the Admission of the State Into the Union*. Chicago: R. R. Donnelley & sons.

Hout, Michael, Clem Brooks, and Jeff Manza. 1995. "The Democratic Class Struggle in the United States, 1948–1992." *American Sociological Review* 60: 805–828.

Huber, Gregory A., and Sanford C. Gordon. 2004. "Accountability and Coercion: Is Justice Blind when It Runs for Office?" *American Political Science Review* 46: 247–263.

Iyer, Lakshmi. 2010. "Direct versus Indirect Colonial Rule in India: Long-term Consequences." *Review of Economics and Statistics* 92: 693–714.

Jackson, John E., Jacek Klich, and Krystyna Poznánska. 2005. *The Political Economy of Poland's Transition: New Firms and Reform Governments*. Cambridge: Cambridge University Press.

Johnson, Amandus. 1930. *The Instruction for Johan Printz, governor of New Sweden: "the first constitution or supreme law of the states of Pennsylvania and Delaware" / translated from the Swedish, with introduction, notes and appendices, including letters from Governor John Winthrop, of Massachusetts, and minutes of courts, sitting in New Sweden, by Amandus Johnson; with a special introduction by John Frederick Lewis*. Philadelphia: Swedish Colonial Society.

Johnson, Harrison. 1880. *Johnson's History of Nebraska*. Omaha: H. Gibson.

Jones, Charles Colcock. 1883. *The History of Georgia*. Boston: Houghton, Mifflin.

Kagan, Robert A., Bliss Cartwright, Lawrence M. Friedman, and Stanton Wheeler. 1978. "The Evolution of State Supreme Courts." *Michigan Law Review* 76: 961–1005.

Karl, T. R., et al. 1986. "A Model to Estimate the Time of Observation Bias Associated with Monthly Mean Maximum, Minimum, and Mean Temperatures for the United States." *Journal of Climate and Applied Meteorology* 25: 145–160.

Karlen, Delmar. 1970. *Judicial Administration: The American Experience*. Dobbs Ferry, N.Y.: Oceana Publications.

Key, V. O. 1949. *Southern Politics in State and Nation*. Reprint, Knoxville: University of Tennessee Press, 1984.

———. 1964. *Politics, Parties, & Pressure Groups*. New York: Crowell.

Kimball, Spencer L. 1966. *Historical Introduction to the Legal System*. St. Paul: West.

King, James D. 1989. "Interparty Competition in the American States: An Examination of Index Components." *Western Political Quarterly* 42: 83–92.

Kinkead, Elizabeth Shelby. 1896. *A History of Kentucky*. New York: American Book Company.

Kleppner, Paul. 1970. *The Cross of Culture: A Social Analysis of Midwestern Politics, 1850–1900*. New York: Free Press.

Klerman, Daniel. 2007. "Jurisdictional Competition and the Evolution of the Common Law." *University of Chicago Law Review* 74: 1179–1226.

———. 2009. "Legal Fictions as Strategic Instruments." Working paper.

Klerman, Daniel, and Paul G. Mahoney. 2005. "The Value of Judicial Independence: Evidence from Eighteenth-Century England." *American Law & Economics Review* 7: 1–27.

———. 2007. "Legal Origin?" *Journal of Comparative Economics* 35: 278–293.

Kopczuk, Wojciech, and Emmanuel Saez. 2004. "Top Wealth Shares in the United States, 1916–2000: Evidence from Estate Tax Returns." NBER Working Paper No. 10399.

Kornhauser, Louis A. 2002. "Is Judicial Independence a Useful Concept?" In Stephen B. Burbank and Barry Friedman, eds., *Judicial Independence at the Crossroads: An Interdisciplinary Approach*. Thousand Oaks, Calif.: Sage.

Kousser, J. Morgan. 1974. *The Shaping of Southern Politics: Suffrage Restriction and the Establishment of the One-Party South, 1880–1910*. New Haven: Yale University Press.

Kruman, Marc W. 1983. *Parties and Politics In North Carolina, 1836–1865*. Baton Rouge: Louisiana State University Press.

La Porta, R., F. Lopez-de-Silanes, C. Pop-Eleches, and A. Shleifer. 2004. "Judicial Checks and Balances." *Journal of Political Economy* 112: 455–470.

La Porta, R., F. Lopez-de-Silanes, and A. Shleifer. 2008. "The Economic Consequences of Legal Origins." *Journal of Economic Literature* 46: 285–332.

LaFountain, R., R. Schauffler, S. Strickland, W. Raftery, C. Bromage, C. Lee, and S. Gibson. 2008. *Examining the Work of State Courts, 2007: A National Perspective from the Court Statistics Project*. National Center for State Courts. Available at http://www.ncsconline.org/D_Research/csp/2007_files/Examining%20Final%20-%202007%20-%201%20-%20Whole%20Doc.pdf.

Lal, Rattan, ed. 1999. *Soil Quality and Soil Erosion*. Ankeny, Iowa: CRC Press.

Landes, William M., and Richard A. Posner. 1975. "The Independent Judiciary as an Interest Group Perspective." *Journal of Law and Economics* 18: 875–902.

Langum, David J. 1987. *Law and Community on the Mexican California Frontier: Anglo-American Expatriates and the Clash of Legal Traditions, 1821–1846*. Norman: University of Oklahoma Press.

Lanman, James Henry. 1841. *History of Michigan: From Its Earliest Colonization to the Present Time*. New York: Harper & Brothers.

Levine, Peter D. 1977. *The Behavior of State Legislative Parties in the Jacksonian Era, New Jersey, 1829–1844*. Rutherford: Fairleigh Dickinson University Press.

Levine, Ross. 2005. "Law, Endowments and Property Rights." *Journal of Economic Perspectives* 19: 61–88.

Lewis, Virgil Anson. 1887. *History of West Virginia*. Philadelphia: Hubbard Bros.

Lieske, J. 1993. "Regional Subcultures of the United States." *Journal of Politics* 55: 888–913.

Liptak, Adam, and Janet Roberts. 2006. "Campaign Cash Mirrors a High Court's Rulings." *New York Times*, October 1.

Lounsberry, Clement Augustus. 1919. *Early History of North Dakota: Essential Outlines of American History*. Washington, D.C.: Liberty Press.

Lowry, Robert, and William H. McCardle. 1891. *A History of Mississippi: From the Discovery of the Great River by Hernando DeSoto, Including the Earliest Settlement Made by the French Under Iberville, to the Death of Jefferson Davis*. Jackson: R.H. Henry.

Lutz, D. S. 1994. "Toward a Theory of Constitutional Amendment." *American Political Science Review* 88: 355–370.

Main, Jackson Turner. 1966. "Government by the People: The American Revolution and the Democratization of the Legislatures." *William and Mary Quarterly* 23: 391–407.

Maine Historical Society. 1919. *Maine: A History*. New York: American Historical Society.

Maizlish, Stephen E. 1983. *The Triumph of Sectionalism: The Transformation of Ohio Politics, 1844–1856*. Kent: Kent State University Press.

Majewski, John, Christopher Baer, and Daniel Klein. 1993. "Responding to Relative Decline: The Plank Road Boom of Antebellum New York." *Journal of Economic History* 53: 106–122.

Malone, Michael P., Richard B. Roeder, and William L. Lang. 1991. *Montana: A History of Two Centuries*. 2nd ed. Seattle: University of Washington Press.

Manchaca, Martha. 2001. *Recovering History, Constructing Race: The Indian, Black, and White Roots of Mexican Americans*. Austin: University of Texas Press.

Manley, Walter W., E. Canter Brown, and Eric W. Rise. 1997. *The Supreme Court of Florida and Its Predecessor Courts, 1821–1917*. Gainesville: University Press of Florida.

Maskin, Eric, and Jean Tirole. 2004. "The Politician and the Judge: Accountability in Government." *American Economic Review* 94: 1034–1054.

Matthews, John Harry. 1987. "Law Enforcement in Spanish East Florida, 1783–1821." Ph.D. dissertation, Catholic University of America.

McConnell, William John. 1913. *Early History of Idaho*. Caldwell: Caxton Printers.

McCormick, Richard P. 1986. *The Party Period and Public Policy: American Politics from the Age of Jackson to the Progressive Era*. New York: Oxford University Press.

McCormick, Richard L. 1974. "Ethno-Cultural Interpretations of Nineteenth-Century American Voting Behavior." *Political Science Quarterly* 89: 351–377.

———. 1979. "The Party Period and Public Policy: An Exploratory Hypothesis." *Journal of American History* 66: 279–298.

McCrady, Edward. 1897. *The History of South Carolina under the Proprietary Government, 1670–1719*. New York: Macmillan.

McGuire, Robert A., and Robert L. Ohsfeldt. 1989. "Self-Interest, Agency Theory, and Political Voting Behavior: The Ratification of the United States Constitution." *American Economic Review* 79: 219–234.

McKnight, Joseph W. 1996. "Spanish Legitim in the United States—Its Survival and Decline." *American Journal of Comparative Law* 44: 75–107.

McSherry, James, and Bartlett Burleigh James. 1904. *History of Maryland*. Baltimore: Baltimore Book Co.

Meany, Edmond Stephen. 1909. *History of the State of Washington*. New York: Macmillan.

Mellinger, Andrew D., Jeffrey D. Sachs, and John L. Gallup. 2000. "Climate, Coastal Proximity, and Development." In Gordon L. Clark, Maryann P. Feldman, and Meric S. Gertler, eds., *Oxford Handbook of Economic Geography*. New York: Oxford University Press.

Merryman, John, and Rogelio Perez-Perdomo. 2007. *The Civil Law Tradition*. 3rd ed. Stanford: Stanford University Press.

Mitchener, Kris James, and Ian W. McLean. 2003. "The Productivity of US States since 1880." *Journal of Economic Growth* 8: 73–114.

Moore, John Wheeler. 1880. *History of North Carolina: From the Earliest Discoveries to the Present Time*. Raleigh: Alfred Williams.

Moreira, M. J. 2003. "A Conditional Likelihood Ratio Test for Structural Models." *Econometrica* 71: 1027–1048.

Morgan, Christopher. 1851. *The Documentary History of the State of New-York*. Albany: Weed, Parsons.

Morphis, J. M. 1875. *History of Texas: From Its Discovery and Settlement, with a Description of Its Principal Cities and Counties, and the Agricultural, Mineral, and Material Resources of the State*. New York: United States Publishing Company.

Motamed, Mesbah J., Raymond J.G.M. Florax, and William A. Masters. 2009. "Geography and Economic Transition: Global Spatial Analysis at the Grid Cell Level." Purdue University. Available at http://www.agecon.purdue.edu/staff/masters/GeoTransRevMar20.pdf.

National Center for State Courts. 2008. State Court Caseload Statistics, 2007. Available at http://www.ncsconline.org/D_Research/csp/2007_files/State%20Court%20Caseload%20Statistics%202007.pdf.

Neatby, Hilda Manion. 1937. *The Administration of Justice under the Quebec Act*. Minneapolis: University of Minnesota Press.

Neill, Edward Duffield. 1858. *The History of Minnesota: From the Earliest French Explorations to the Present Time*. Philadelphia: J.B. Lippincott.

Nelson, William E. 1982. *The Roots of American Bureaucracy, 1830–1900*. Cambridge: Harvard University Press.

———. 2000. *Marbury v. Madison: The Origins and Legacy of Judicial Review*. Lawrence: University Press of Kansas.

———. 2008. *The Common Law of Colonial America*. Vol. 1, *The Chesapeake and New England, 1607–1660*. New York: Oxford University Press.

North, Douglass C. 1966. *The Economic Growth of the United States: 1790–1860*. New York: Norton.

———. 1981. *Structure and Change in Economic History*. New York: Norton.

———. 1990. *Institutions, Institutional Change and Economic Performance*. New York: Cambridge University Press.

Nunn, Nathan, and Diego Puga. 2009. "Ruggednesss: The Blessing of Bad Geography in Africa." NBER Working Paper No. 14918.

O'Callaghan, Edmund Bailey. 1848. *History of New Netherland or, New York Under the Dutch*. New York: D. Appleton.

Olson, James C., and Ronald Clinton Naugle. 1997. *History of Nebraska*. 3rd ed. Lincoln: University of Nebraska Press.

Ostrom, Elinor. 1990. *Governing the Commons: The Evolution of Institutions for Collective Action*. New York: Cambridge University Press.

Parkins, Almon Ernest. 1918. *The Historical Geography of Detroit*. Lansing: Michigan Historical Commission.

Patterson, Samuel, and Gregory Caldeira. 1984. "The Etiology of Partisan Competition." *American Political Science Review* 78: 691–707.

Perkins, James Handasyd, John Mason Peck, and James R. Albach. 1854. *Annals of the West*. St. Louis: James R. Albach.

Pessen, Edward. 1973. *Riches, Class, and Power before the Civil War*. New York: D. C. Heath.

———. 1980. "How Different from Each Other Were the Antebellum North and South?" *American Historical Review* 85: 1119–1149.

Phelan, James. 1888. *History of Tennessee: The Making of a State*. Boston: Houghton, Mifflin.

Pickett, Albert James, and Thomas McAdory Owen. 1900. *History of Alabama and Incidentally of Georgia and Mississippi, from the Earliest Period: From the Earliest Period*. Birmingham: Webb Book Company.

Poole, Keith T., and Howard Rosenthal. 1997. *Congress: A Political-Economic History of Roll Call Voting*. New York: Oxford University Press.

Pound, Roscoe. 1906. 29 *A.B.A. Rep.* 395, 410–411. Reprinted in *Baylor Law Review* 8 (1956): 1–25.

Prince, Le Baron Bradford. 1912. *A Concise History of New Mexico*. Cedar Rapids, Iowa: Torch Press.

Purvis, Thomas L. 1984. "The European Ancestry of the United States Population, 1790." *William and Mary Quarterly* 41: 85–101.

Ramsay, David. 1809. *The History of South-Carolina: From Its First Settlement in 1670, to the Year 1808*. Charleston: David Longworth.

Ramseyer, J. Mark. 1994. "The Puzzling (In)Dependence of Courts: A Comparative Approach." *Journal of Legal Studies* 23: 721–747.

Rappaport, Jordan, and Jeffrey D. Sachs. 2003. "The United States as a Coastal Nation." *Journal of Economic Growth* 8: 5–46.

Rees, Albert. 1975. *Real Wages in Manufacturing, 1890–1914*. New York: Arno Press.

Remington, Thomas. 2011. *The Politics of Inequality in the Russian Regions*. Forthcoming from Cambridge University Press.

Report of the Public Lands Commission. 1905. Washington, D.C.: Government Printing Office.

Reynolds, John. 1887. *The Pioneer History of Illinois: Containing the Discovery, in 1673, and the History of the Country to the Year 1818, when the State Government was Organized*. Chicago: Fergus Printing Company.

Rink, Oliver A. 1986. *Holland on the Hudson: An Economic and Social History of Dutch New York*. Ithaca, N.Y.: Cornell University Press.

Robinson, Doane. 1905. *A Brief History of South Dakota*. New York: American Book Co.

Rodrik, Dani. 1999. "Democracies Pay Higher Wages." *Quarterly Journal of Economics* 114: 707–738.

Rosen, Deborah A. 2001. "*Acoma v. Laguna* and the Transition from Spanish Colonial Law to American Civil Procedure in New Mexico." *Law and History Review* 19: 513–546.

Rosenburg, Gerald N. 1991. *The Hollow Hope: Can Courts Bring About Social Change?* Chicago: University of Chicago Press.

Ross, Michael. 2001. "Does Oil Hinder Democracy?" *World Politics* 53: 325–361.

Sachs, Jeffrey D. 2003. "Institutions Don't Rule: Direct Effects of Geography on Per Capita Income." NBER Working Paper No. 9490, February.

Sachs, Jeffrey D., and Pia Malaney. 2002. "The Economic and Social Burden of Malaria." *Nature* 415: 680–685.

Sachs, Jeffrey D., and Andrew Warner. 1997. "Natural Resource Abundance and Economic Growth." CID Working Paper.

Sage, Leland Livingston. 1974. *A History of Iowa*. Ames: Iowa State University Press.

Schaefer, Donald. 1983. "The Effect of the 1859 Crop Year Upon Relative Productivity in the Antebellum Cotton South." *Journal of Economic History* 43: 851–865.

Scharf, John Thomas. 1888. *History of Delaware, 1609–1888*. Philadelphia: L. J. Richards.

Schell, Herbert Samuel. 1968. *History of South Dakota*. 2nd ed. Lincoln: University of Nebraska Press.

Sharkansky, Ira. 1969. "The Utility of Elazar's Politic Culture: A Research Note." *Polity* 2: 66–83.

Sharpless, Isaac. 1900. *Two Centuries of Pennsylvania History*. Philadelphia: J.B. Lippincott.

Shepherd, Joanna. 2009. "The Influence of Retention Politics on Judges' Voting." *Journal of Legal Studies* 38: 169–203.

Sheridan, Thomas E. 1995. *Arizona: A History*. Tucson: University of Arizona Press.

Shugerman, Jed. 2009. "The Twist of Long Terms: Disasters, Elected Judges and American Tort Law." *Georgetown Law Journal* 98: 1349–1414.

Smith, Adam. 1914. *An Inquiry into the Nature and Causes of the Wealth of Nations*. New York: E. P. Dutton.

Smith, Samuel. 1877. *The History of the Colony of Nova Cæsaria, Or New Jersey: Containing, an Account of Its First Settlement, Progressive Improvements, the Original and Present Constitution, and Other Events to the Year 1721. With Some Particulars Since and a Short View of Its Present State*. Trenton: W.S. Sharp.

Smith, William Rudolph. 1854. *The History of Wisconsin*. Madison: B. Brown,

Snook, James Eugene. 1904. *Colorado History and Government, with State Constitution*. Denver: Herrick Book and Stationery Company.

Snowden, Clinton A. 1909. *History of Washington: The Rise and Progress of an American State*. Advisory editors Cornelius Holgate Hanford, Miles C. Moore, William D. Tyler, and Stephen J. Chadwick. New York: Century History Company.

Sokoloff, Kenneth L., and Stanley L. Engerman. 2000. "History Lessons: Institutions, Factor Endowments, and Paths of Development in the New World." *Journal of Economic Perspectives* 14: 217–232.

Soltow, Lee. 1975. *Men and Wealth in the United States, 1850–1870*. New Haven: Yale University Press.

———. 1989. *Distribution of Wealth and Income in the United States in 1798*. Pittsburgh: University of Pittsburgh Press.

Sommeiller, Estelle. 2006. "Regional Income Inequality in the United States, 1913–2003." Ph.D dissertation, Department of Economics, University of Delaware.

Squire, Peverill. 2006. "The Professionalization of State Legislatures in the United States Over the Last Century." Revised version of paper presented at the 20th International Political Science Association World Congress, Fukoka, Japan, July.

———. 2007. "Measuring Legislative Professionalism: The Squire Index Revisited." *Politics and Policy Quarterly* 7: 211–228.

Squire, Peverill, and Keith Hamm. 2005. *101 Chambers: Congress, State Legislatures, and the Future of Legislative Studies*. Columbus: Ohio State University Press.

Starr, Kevin. 2005. *California: A History*. New York: Random House.

The State Register and Year Book of Facts. 1859. San Francisco: Henry G. Langley and Samuel A. Morrison.

Steckel, R. H., and C. M. Moehling. 2001. "Rising Inequality: Trends in the Distribution of Wealth in Industrializing New England." *Journal of Economic History* 61: 160–183.

Stith, William. 1865. *The History of the First Discovery and Settlement of Virginia*. New York: Joseph Sabin.

Stone Sweet, Alec. 1992. *The Birth of Judicial Politics in France*. New York: Oxford University Press.

Stout, Wayne Dunham. 1971. *History of Utah*. Salt Lake City.

Sturm, Albert. 1981. "The Development of American State Constitutions." *Publius* 12:57–98.

Tabarrok, Alexander, and Eric Helland. 1999. "Court Politics: The Political Economy of Tort Awards." *Journal of Law and Economics* 42: 157–188.

Tamahana, Brian. 2006. *Law as a Means to an End: Threat to the Rule of Law*. Cambridge: Cambridge University Press.

Tarr, G. Alan. 1992. "Understanding State Constitutions." *Temple Law Review* 65: 1169–1198.

———. 1996. *Constitutional Politics in the States: Contemporary Controversies and Historical Patterns*. Westport, Conn.: Greenwood Press.

———. 2000. *Understanding State Constitutions*. Princeton: Princeton University Press.

Taylor, George Rogers. 1951. *The Transportation Revolution, 1815–1860*. New York: Rinehart.

Taylor, Stuart, Jr. 1988. "High Court Expands States' Rights to Tidelands Miles from the Ocean." *New York Times*, February 24.

Tebeau, Charlton W. 1972. *A History of Florida*. Coral Gables: University of Miami Press.

Thoburn, Joseph Bradfield, and Isaac Mason Holcomb. 1908. *A History of Oklahoma*. San Francisco: Doub.

Tobin, Robert W. 1999. *Creating the Judicial Branch: The Unfinished Reform*. Williamsburg, Va.: National Center for State Courts.

Trumbull, Benjamin. 1898. *A Complete History of Connecticut, Civil and Ecclesiastical: From the Emigration of Its First Planters, from England, in the Year 1630, to the Year 1764, and to the Close of the Indian Wars*. New London: Utley.

Tuttle, Charles Richard. 1875. *An Illustrated History of the State of Wisconsin: Being a Complete Civil, Political, and Military History of the State, from Its First Exploration Down to 1875*. Madison: B.B. Russell.

Tyler, Gus. 1962. "Court vs. Legislature (The Sociopolitics of Malapportionment)." *Law and Contemporary Problems* 27: 390–407.

Ubbelohde, Carl, Duane A. Smith, and Maxine Benson. 2006. *A Colorado History*. 9th ed. Boulder: Pruett.

U.S. Census Bureau. 2001. *Federal, State and Local Governments: Government Finance and Employment Classification Manual*.

van Kleffens, Eelco Nicolaas. 1968. *Hispanic Law Until the End of the Middle Ages: With a Note on the Continued Validity After the Fifteenth Century of Medieval Hispanic Legislation in Spain, the Americas, Asia, and Africa*. Edinburgh: Edinburgh University Press.

Van Laer, Arnold J. F. 1974. *Register of the Provincial Secretary; Council Minutes*. 4 vols. Baltimore: Genealogical Publishing.

Vigil, Maurilio E. 1980. *Los Patrones: Profiles of Hispanic Political Leaders in New Mexico History*. Washington, D.C.: University Press of America.

Vigneras, Louis-Andre. 1969. "A Spanish Discovery of North Carolina in 1556." *North Carolina Historical Review* 46: 398–407.

Vincent, Francis. 1870. *A History of the State of Delaware: From Its First Settlement Until the Present Time, Containing a Full Account of the First Dutch and Swedish Settlements, with a Description of Its Geography and Geology*. Philadelphia: J. Campbell.

Violette, Eugene Morrow. 1918. *A History of Missouri*. New York: D. C. Heath.

Volcansek, Mary, Maria De Franciscis, and Jacqueline Lafon. 1996. *Judicial Misconduct: A Cross-National Comparison*. Gainesville: University Press of Florida.

Wagoner, Jay J. 1970. *Arizona Territory 1863–1912: A Political history*. Tucson: University of Arizona Press.

Wallace, Joseph. 1893. *The History of Illinois and Louisiana under the French Rule: Embracing a General View of the French Dominion in North America with Some Account of the English Occupation of Illinois*. Cincinnati: R. Clarke.

Wallis, John J. 2005. "Constitutions, Corporations, and Corruption: American States and Constitutional Change, 1842 to 1852." *Journal of Economic History* 65: 211–256.

———. 2006. "The Concept of Systemic Corruption in American History." In Edward Glaeser and Claudia Goldin, eds., *Corruption and Reform: Lessons from America's Economic History*. Chicago: University of Chicago Press.

———. 2008. "The Other Foundings: Federalism and the Constitutional Structure of American Government." Working paper.

Watson, Harry L. 1997. "Humbug? Bah! Altschuler and Blumin and the Riddle of the Antebellum Electorate." *Journal of American History* 84: 886–893.

Weber, David J. 1982. *The Mexican Frontier, 1821–1846: The American Southwest under Mexico*. Albuquerque: University of New Mexico Press.

Wiener, Jonathan. 1975. "Planter-Merchant Conflict in Reconstruction Alabama." *Past and Present* 68: 73–94.

Wilentz, Sean. 1982. "On Class and Politics in Jacksonian America." *Reviews in American History* 10: 45–63.

Wilkins, Robert Poole, and Wynona H. Wilkins. 1977. *North Dakota: A Bicentennial History*. New York: Norton.

Woodman, Harold. 1962. "Chicago Businessmen and the Granger Laws." *Agricultural History* 36: 16–24.

Wooster, Ralph. 1969. *The People in Power: Courthouse and Statehouse in the Lower South, 1850–1860*. Knoxville: University of Tennessee Press.

———. 1975. *Politicians, Planters, and Plain Folk: Courthouse and Statehouse in the Upper South, 1850–1860*. Knoxville: University of Tennessee Press.

Worcester, Donald E. 1976. "The Significance of the Spanish Borderlands to the United States." *Western Historical Quarterly* 7: 5–8.

Young, Crawford. 1994. *The African Colonial State in a Comparative Perspective*. New Haven: Yale University Press.

Zweigert, Konrad, and Hein Kötz. 1998. *An Introduction to Comparative Law*. 3rd ed. New York: Oxford University Press.

Index